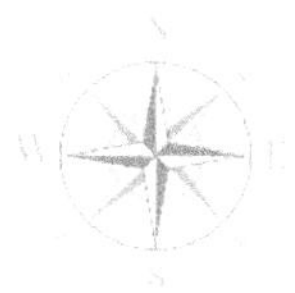

"What road do I take?"

"Well, where are you going?

"I don't know..."

"Then it doesn't matter.

If you don't know where you are going,

any road will get you there."

~ Alice in Wonderland

Pamdiana Jones

When in Roam

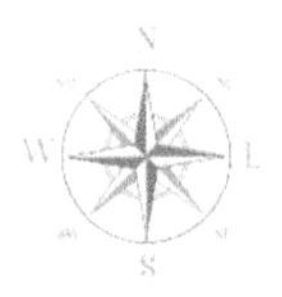

Author's Note

My warmest heartfelt Thanks to everyone reading about my wacky adventures.

Enjoy all my clumsy, cringeworthy, naïve, scary, half-offensive, funny, awkward moments... all written for the different time we were living in the mid 1990's.

WARNING: Please enter my R-rated story at your own risk. This story contains controversial moments, yet I simply must tell my genuine account.

Although this is a true story, some of the names have been changed to protect the shameless, the wild, and the guilty. Events, timelines, and conversations are written as accurately as I can recall, they are not intended to be exact.

I offer my deepest gratitude to all who shared their homes, their ideas, their lives, and their laughter with me along my travels.

Cover Art by *Bethany Barnett*

Cover Art and Travel Maps by *Dawn Teagarden*

Turtle Publishing House

ISBN: 978-1-7352736-00

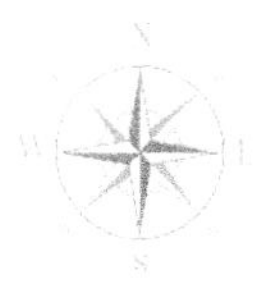

Dedicated to my family

...for my twins

son Bryce Harrison
daughter Hunter Ferial

who may read this book once they turn 30!

...for my husband

Daniel
who loves me for my adventurous nature...

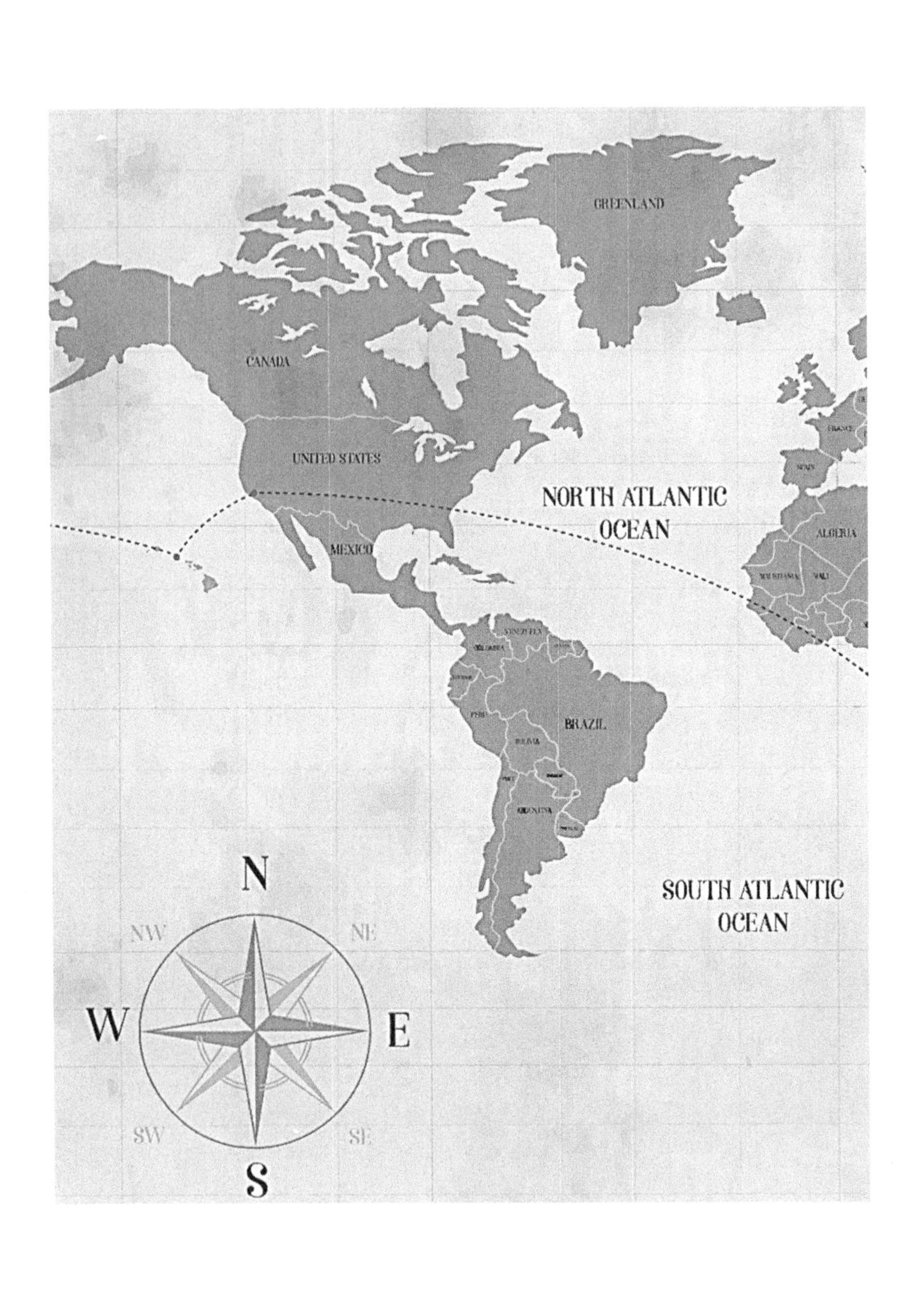
GREENLAND
CANADA
UNITED STATES
NORTH ATLANTIC
OCEAN
MEXICO
SPAIN
ALGERIA
BRAZIL
SOUTH ATLANTIC
OCEAN
N
NW
NE
W
E
SW
SE
S

ARCTIC OCEAN
RUSSIA
POLAND
KAZAKHSTAN
MONGOLIA
TURKEY
CHINA
PAKISTAN
INDIA
PACIFIC OCEAN
LIBYA
EGYPT
CHAD
SUDAN
KENYA
INDONESIA
INDIAN OCEAN
AUSTRALIA

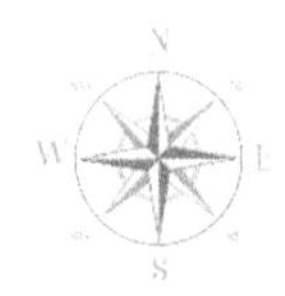

Table of Contents

Foreword

Rod Stewart changed my life and he doesn't even remember me.

At 19 years old, I was living in the outskirts of Los Angeles—*my first clue*—waiting tables at a fancy restaurant—*yep, second*. As I greeted the two men in my section and took their drink order, I noticed one of them looked a lot like the singer Rod Stewart.

As I placed their order with the bartender, he asked, "Are you waiting on Rod Stewart?"

I shook my head, "No, he just looks a little like him," I said.

"Oh," he frowned, handing me the drinks.

I delivered the drinks and took their lunch order. In the kitchen, I saw the hostess talking with the owner and the chef. They asked me the same thing, "Is that Rod Stewart at table eight?"

"No, just a guy who styled his hair like him," I said with confidence.

"Oh..." they said, disappointment clear in their voices.

I continued waiting on other tables and a few of them asked me as well, "Hey Miss, is that Rod Stewart over at that table?"

"No," I smiled, and realized they could just see that spiky haircut over the top of the booth.

"Oh," clearly, they thought I was protecting his privacy so he could enjoy his lunch in peace.

As I began clearing the plates, Mr. Spiked Hair looked up at me and I heard the British accent, "Hey Darling, would you like to be in a video?"

Oh brother, he's really over the top.

He's really working this 'I look like Rod Stewart and use it to pick up chicks' *thing.*

Could I trust this guy?

"Me? In a video?" I was amused.

"Yes, we're making a music video. Would you like to be in it?" he tried again.

I steadied my hands full of dirty dishes.

A huge grin spread across my face...

*****

South Africa

Johannesburg

- Tonight's the Night -
Rod Stewart

I'm lost in my daydream.

Who will I meet? What adventures will I have? Who will I become?

Here comes that loud, rumbling, clickety-clack noise again, snapping me back to reality. I steady my legs, moving with the rhythm.

Through the dusty train window, I catch my reflection flickering in the passing train's glass. My gauzy, lemon-yellow tunic shirt. Vintage, borrowed from my mom's travels in the 70's. My long, wavy, auburn hair, unbrushed and blowing back in the wind. My cinnamon brown eyes bloodshot, seared with exhaustion, yet still gleaming with excitement. Smudged eyeliner.

I still look like me... only wilder somehow, untamed already.

After a 29-hour flight from Los Angeles and *yep—that's not including stopovers—* my friend J.J. (Jay) and I had been on the train for 28 hours from Johannesburg to Cape Town, South Africa.

Throughout the picturesque journey, we'd passed living, breathing postcards. Misty morning fog nestled into the bases of towering shale rock mountains. Golden sunflower fields with babbling brooks. Rows of climbing violet and green grape vines at the wineries. Modest country apple orchards. Wide open meadows. Rusty tin shanty-towns, covered in graffiti and razor wire, a river of raw sewage running down the dirt street. *Maybe I'll skip that last postcard...*

It finally hits me.

I am in AFRICA.

My final frontier.

It's springtime in 1995, I'm 24 years old. Let the adventure begin.

After my train's tiny cubicle hot shower, I see the city. It looks like we're pulling into the station momentarily. I run down the length of the cars until I reach the restaurant car, plopping down next to J.J. and our new pals, Arthur and Samson. Readying my backpack, I'm excited to begin exploring.

They all seem serious, even concerned. The train is slowing.

"Enzo stabbed someone!" Jay reveals.

Who the hell is Enzo?

*****

I think back to how it all began.

I grew up north of Los Angeles, California, in Santa Clarita Valley. Home of the Magic Mountain theme park, where Bugs Bunny characters reside. Middle class with good friends and a loving family, my parents hit a rough patch and were divorcing. Restless and searching for adventure, I'd moved out at 18 years old. It was spontaneous, to say the least.

"What do you feel like doing today?" a friend had asked me.

"Let's drive to Florida," I'd cracked a joke.

So, we did. I left my mom a yellow Post-It note saying '*I moved to Florida. I'll call you from ??? tomorrow!*' The note did not go over well with my parents, but still, I was gone for six months. We ended up in South Carolina for Hurricane Hugo, where I worked as a roofer with my friends, and Daytona Beach, Florida, for MTV's Spring Break.

I couldn't settle in back at home and moved around a few times. First to San Diego, where I was devastated when my moving truck full of everything I'd owned had been stolen. My entire childhood, my entire identity, *gone*.

Then, to Park City, Utah, with my sister Sandy. We waited tables and snowboarded in the ski resort town. That's where I'd met J.J.

J.J. grew up south of LA, in nearby Huntington Beach. He was a California surfer with his tanned skin, lean muscles, and long ponytail bleached blond from the salt water. He was the epitome of the '*Fast Times at Ridgemont High*' Spicoli character. He'd just moved to Utah to save up for an African surf trip.

"That sounds AWESOME..." I said, in, well... *awe*.

"The ticket's $1,175 if you want to go?" Jay mentioned casually.

Hmmm, I have exactly $1,200 in the bank right now, and if I sell my car for $3,000, this might actually work. "I'm in!" I committed without a second thought. I emptied my bank account and brought the money the next day.

When I handed the cash over to Jay, he was shocked.

"Everyone says they're up for it, but then have all these excuses why they can't go," he said. We had a month before we left, which gave me time to sell my car. I ended up with $3,300 to travel three months in South Africa with Jay, then he would continue on with a stopover in Indonesia, while I would head home after Africa.

Before we left, he advised me to get a small backpack, not one I couldn't carry, and definitely not one that advertised to thieves, "*Hey, I'm far from home with everything I own on me!*" I'd bought a navy-blue schoolbook bag with small side pouches that could detach for a hiking day. I'd packed one pair of jeans, one long-sleeve shirt, a few T-shirts, tank tops, a sarong, a blue bikini, three pairs of undies, two pairs of socks, one bra, a summer dress, flip-flops, hiking boots, and some cosmetics. My mom had given me a travel teddy bear, with a fluffy beach towel that took up half the bag.

Our first day together, J.J. had pulled out of his school-sized backpack a thin hand towel to use as a proper shower towel. Along with his clothes, he had some rope, a Swiss army knife, safety pins, needle and thread, a small flashlight, a journal, a map, and a compass. *Ahhhh yes, MacGyver. That makes way more sense.* And, of course, his surfboard.

After landing in Johannesburg, I wouldn't have even figured out how to leave the airport but for Jay. *Hey, someone had always picked me up.* Poor J.J., I hoped he'd keep me from making a fool of myself, since I'm a *new* traveler. Unwrapped Christmas present *new*. Fresh snow with no footprints *new*. He'd traveled quite a bit in his 21 years, but this was eye-opening for me.

Obviously, Jay taught me to go to the Visitor's Information booth. We found where to catch a bus and changed our US Dollars into South African Rand. *Travel weary* and *jetlag* became my new best friends. We headed to a hostel and slept for 20 hours straight.

I'd never even heard of a hostel before. They're cheap accommodations with about six bunkbeds to each guest room, with dormitory restrooms and showers at the end of the hall. The rec room had murky, mismatched couches, a small TV, even a few books and games. Indian curry and motor oil wafted through the hallways. More razor wire fences to deter burglars. Our room overlooked the luxury infinity-edge glass pool. *I mean the dark-green, algae-filled, shoebox of a slime pit. Not exactly the Ritz.*

After stopping at a mini market, we stored our own food in the fridge and cooked for ourselves in the cafeteria kitchen.

Over a few hours the next day, we tried and tried to buy a train ticket out of Jo'burg in either first or second-class but were denied. They'd sold out for the next month due to school holidays.

Just the thought of a landlocked month in a barbed-wire and metal bars-on-every-window kind of city, one that we learned was *20-times more violent than New York City*, made us both antsy. We had to get out of this stifling city and head to the beach.

At first light, we marched back to the train station. Jay slapped his hair into a ponytail, his normally warm brown eyes fierce with new determination. We insisted on third-class tickets to the handlebar-mustached ticket agent who'd denied us the day before. He could tell we were strong-willed, yet clueless tourists who didn't know what we were getting into. He shook his head, frowning.

"It's full of only blacks... they *will* hassle her," he said as he looked me up and down.

"Isn't that reverse discrimination?" Jay asked.

Still shaking his head in disbelief, he twirled the long twist of his waxed 'stache. We shoved the money through the window. He sighed and reluctantly handed us the tickets.

I was quite proud of us, I'll admit it, even cocky. Like "*I read the newspapers*" and "*I know all about your apartheid*" and "*We'll show these African black families how well we respect their culture*" and even "*I'm from LA, and I have lots of black friends!*"

I was instantly humbled as I stepped into the third-class car, where hundreds of dark faces stared at us in utter disbelief. We just took third-class to have an adventure. These families had no choice but to take it.

There was small table-and-booth seating, like RV camping. The units seated four people comfortably, but each one was overstuffed with about six, seven, even eight people. An armed security officer stood just outside the sliding train door, in case of emergency. At just 7:00 a.m. it was 85 degrees Fahrenheit, but every last person was wearing a Bill Cosby sweater with *Every. Single. Window. Rolled. Up.*

The BO, the booze, the litter, the bad breath... and I was positive there were dirty diapers that needed to be changed. Immediately.

With the only white faces in the group, we smiled, nodded our heads, and said hello as we stuffed our two bags overhead. After asking politely, we squeezed in. *Hey, what's that saying? 'When in Rome, you do as the Romans do.'* Over the next few hours, we made small talk and tried to roll down a window now and then. They smiled at us and rolled it back up. We smiled and rolled it down a bit later, and again, they smiled and rolled it back up. I tried to catch the eye of a couple of women nearby. I wanted to say hello, but they deliberately avoided eye contact.

The air was beyond stuffy. It was nearly winter here in the southern hemisphere, but at least 90 degrees. I just couldn't breathe. I was lightheaded, fidgety. Our sweaty arms stuck together, I leaned into Jay and whispered that I *might* need to throw up. He grabbed my arm tight, frowning intensely.

"Do NOT do it in here!" he hissed.

My heart skipped a beat at the seriousness of his tone. I quickly headed to the bathroom. Upon seeing the third-class facilities, I even more quickly headed to the air-conditioned, first-class dining car restrooms. *Well, Rome certainly had a first class as well, thank you very much.*

The moment passed after I splashed cold water on my face. As I meandered back through the dining car, two men smiled at me, waving me over to their table.

"Hello, how's it? I'm Arthur and this is Samson," Arthur said, in his early 30's with a receding, sand-colored mullet and a genuine smile. He stood as he shook my hand.

"Please, won't you join us for a moment?" asked Samson, maybe in his 50's, a little chubby, with salt and pepper hair.

"Hello, I'm Pam. Thank you, yes, of course I'll join you," I said, happy to hear their South African accents.

"We saw you and your friend get on in third-class back in Jo'burg," Samson said. "We were just about ready to come back and look for you both as it gets quite wild back there at nighttime."

Trying to sound tough, I smiled, "I *think* we'll be okay..."

After some more chatting, these men offered to buy us dinner. Wow. They knew I was traveling with a man, so this wasn't a pick-up line. What kind, generous people. Happy to be more comfortable, I ran back down the train to tell Jay.

As I returned to third-class, the stench was upon me again and I found Jay asleep with another man's head on his shoulder. *Aww, cuddle buddies.* I tried to wake Jay, but he was unresponsive. I called him and he didn't move. Comical thoughts vanished as I wondered if he'd been drugged. I shook him hard, yelling, "Jay, wake up!" until he groggily gained his bearings.

"I've met a couple of great guys who invited us to dinner. They said to grab our bags and come on up." Jay asked me to help get his bag down, quietly hinting that his money belt would've been visible if his shirt lifted when he reached overhead. *Hmmm, I'd better learn to think that way. Like a traveler.*

After the introductions, Arthur said they'd been police officers together. Samson had retired, and they'd started a car delivery service. They'd drive a couple of cars from Cape Town to Jo'burg and take the train back every few days. They traveled with a few workers they called *coloured people* (of perhaps Indian descent, Malay, or maybe Islanders), who weren't allowed in the first-class dining room.

Seriously? Still doing that?

Conversation flowed easily, and after asking us all about America, they explained more of the problems in South Africa and how nightfall in third-class was beyond dangerous. That security guard outside the doors all afternoon? He'd lock the doors, go off to bed, and that's when all hell would break loose. Every single night someone got *stabbed.*

"What about shot?" Jay inquired.

"No, no guns. Only the coppers and random militia groups have guns. Lots of stabbings, everyone gets stabbed," Samson replied, readying his napkin.

Jay and I looked at each other and... *smiled.* We were both thinking it. *Sweet! No guns. Bad guys must get closer to stab someone. Unexpectedly awesome.*

Our rosemary chicken dinners arrived and the smell was mouth-wateringly delicious. Despite the last word spoken being *stabbed*, I stabbed at my chicken, starved for that first bite.

"But there are women and children in there!" I cried in disbelief, sipping from the red wine, my chipmunk cheeks filled. *Manners, schmanners, hey, I had a bowl of cereal eight hours ago.*

"They will turn away as you are getting raped and stabbed, even skinned alive," Arthur replied as he skinned his chicken. I thought back to that ticket agent. "*They will hassle her,*" he'd said. If I'd known that '*hassle*' meant '*raped and skinned alive,*' I never would've agreed to a third-class ticket.

Samson continued, "But we are not just buying dinner, we require you sleep in our second-class cabin with us. Our two coloured workers will take your places in the back, and they know how to deal with situations that might come up. They will be perfectly fine. We absolutely insist."

Well, after that scare, how could we say no?

We were led to the cabin car and got comfortable lying down on bunk beds. No blankets or pillows. Everyone snored the night away but me. The wind was whipping through the crack in the window, going right through my boots, and even through my damp socks I'd been wearing unwashed for three days now. My feet were so cold I could barely brave it. It felt like little frozen nails hammering into my toes all night.

My worries also kept me up. *What if we'd caught the train the day before?* I couldn't even think about it. It's no longer just a fun adventure. No matter what I think about racism, prejudice, and my own country's diversities, I must admit it... I know nothing about what it's really like here and I'd better zip my lip, check my attitude, and do what other white people do to stay safe.

Morning arrived and we enjoyed a champagne breakfast paid for by our hosts. These guys rock.

We talked further about the apartheid that ended last year. The word apartheid literally means '*apartness*' in Afrikaans. Raised to hate each other's races, yet in just one year, expected to be friends. If you were a black maid or chauffeur for a wealthy white family, politicians promised if you voted for Nelson Mandela, it would become *your* new house and *your* new limousine. There'd been TV commercials urging, "*Go kill the family you work for and take what belongs to you.*" At the one-year celebration of the election, people waved their new South African flags, yet had scowls on their faces. A year after the election, those promises were proven false.

I think I can understand a fraction of that anger. Since just last night, I'd experienced a free meal, a comfy cabin to sleep in, and a free champagne brunch, *only because I'm white.* No one offered that to any black families

*****

Back in the now, finished with my train shower, I've just caught my reflection in the glass. I'm done catching up on my reverie and daydreams. I'm fully present as I sit with Jay, Samson, and Arthur.

"Enzo stabbed someone!" Jay says.

The train lets out a steamy sigh as we come full stop into the station.

"What? Is everyone all right?" *Who? Is this good news or bad news?* I don't even know what to ask.

Stepping off the train in Cape Town, we follow Arthur and Samson to finally meet Enzo and Dian, the two gentlemen who took our spots in third-class. We learn Enzo *had to stab someone while he was in our place.*

My jaw drops as he tells his tale of how a huge bald black man had approached him with a switchblade to rob him. Enzo, a small man of maybe 5'4", had to wrestle the knife in his attackers' hands until the knife was upside down and facing back toward the criminal. *He made the huge man stab himself.* Then, as the bad guy crawled back to his family, bleeding all over the train floor, our two saviors had to act tough for the rest of the night so no one else would attack them as they slept. They rode the rest of the way on their feet, standing back to back.

But, yeah, my toes were chilly.

Raw shame washes over me that we'd so quickly accepted the switch, thinking only of ourselves. I can't look them in the eye. The fear these men had gone through, strictly because of us. *I hadn't even known their names.*

Mortified, we ask Samson and Arthur if it would be appropriate to compensate them for their horrific night. A pack of cigarettes is all our Angels from Heaven will accept.

I find it unbelievable that our choices could lead to life and death for others and their families. Even the bad guy... *is he dead?* Is he a husband? A father? The train hadn't stopped in the middle of the night to allow him a trip to the hospital.

Chaos Theory is in full effect: If a butterfly flaps its wings off the coast of Morocco, it somehow causes a tornado in Idaho a year later that wipes out a town. Similar in effect, but on a much more personal level.

Jay and I steal a glance at each other.

It would've been a grossly different outcome if we'd been back there. The end for us both.

I've been a traveler for exactly three days in Africa and the weight of this night has aged me.

Cape Town

- Waiting for the Miracle -
Leonard Cohen

Nestled between the flat Table Mountain—with its "tablecloth" of cloud coverage— and stunning beaches, sits the quaint city of Cape Town. Its architecture reminds me of New Orleans. My grandparents had lived as British expats in Africa for 10 years in the 1930's and my mother had traveled to Kenya and Tanzania in the 70's. I'd grown up hearing their stories and can't believe I'm really here. And even mom's yellow shirt had survived 20 years to return home to the motherland with me.

We settle into the Long Street Backpackers hostel, get a small room with twin beds. We meet a mix of new friends, mostly Europeans and Australians ready to drink the night away. But one man interests me. A black man introduces himself to me, smiling.

"Hello. I am Ken from Kenya." Shaking hands, he continues in a deep voice, "You are the very first person in my life to just say 'nice to meet you' not making fun that I am Ken from Kenya."

We laugh together, and I share with him my mother's stories about her trip to Kenya. He left his country a year ago to escape some violence and has been traveling ever since. He's warm and funny. He asks all about America. We talk for ages.

After the fear had been put into me of this different culture on the train, I really needed this balance to remind myself that people are people wherever I go.

In the first few days of traveling, I've spent an entire day flying, an entire day sleeping, and a third full day on a train. It's definitely laundry time.

"Hey, I stink. Should we look for a laundromat? I'm starving too, maybe we could find a café?" I ask.

Jay teaches me the ropes of traveling in one sentence. "You know, the least amount of money we spend means we can keep traveling longer."

Okay, what does that mean?

He continues, "Instead of spending even just $5 at a laundromat, why not wash our clothes in the sink with a bar of soap? $5 at a fast food place, each meal adds up quickly. Why not buy a bag of white rice for $2 and that will last us all week for breakfast, lunch, and dinner? $5 is an entire day in Indonesia, a hut, three meals, and even a taxi ride. I'm heading there on the way back from here, so each dollar has to be worth it."

But white rice every day? *Every meal*?

"Do you put, like..." I grimace, my face distorted, "...butter or something on it?" *But I'm up-for-anything-girl. Sigh.*

I finally get what we're taking on. White rice. Breakfast, lunch, and dinner for three months is going to be tough for me. I'm a foodie and a wino. But after long years of constantly working, we're taking a break from that life, and there is exactly zero money coming in. We have to conserve to make it last. *Duh.*

Over the next few days in Cape Town, J.J. and I explore the city. He's starting to grow a tiny thin beard and just trims it regularly with nail scissors to save money on razors. He refuses to watch TV, stating it turns people's minds to mush. We check out the bookstore, tourist information, the daily flea market around the corner. We hike up to Lion's Head Mountain where we find a tire swing and enjoy the view of the sea. We plan to hike up Table Mountain when the clouds move on. The days are sunny and warm, but the nights are freezing. We listen to our Walkman tapes while lying by the pool. We wander the supermarkets. It's fun to see the different brands and logos, even if we don't buy anything.

We even answer an ad to purchase a car. The sellers give us camping equipment: pots and pans, metal plates and mugs, a map of the country, and they even mark cool spots to stop along the coast. Our new baby, a Ford Cortina we proudly name *Tina*, is definitely *not* road worthy, but cost just $700US. She's a small red station wagon that plays only classical music. *Haunted car 'Christine'*? The front doors don't lock properly, so we attach a chain that hooks the doors together permanently. We have to clamber in and out through the windows. *'Dukes of Hazzard'*?

It does feel great to be mobile again. Out of habit, I accidentally keep walking up on the same side of the car as Jay each time we get into it. I swear, I'm not doing it on purpose. They just drive on the wrong side of the road.

"Hey," I ask, once we're on the open road for an hour, "Do they have tow trucks in Africa?"

Jay just rolls his eyes.

We get into our groove. We buy a thick piece of foam to place in the back of the car so we can sleep in the hatch under the stars. Over the next week, we camp down at Cape of Good Hope, the southern tip of Africa, where the two oceans meet. Jay surfs, and we drive through Simonstown, a fishing village with pastel-colored buildings. There are even penguins at the nearby beach. Who knew? We buy a second surfboard, a longer blue one I can try in the warmer water as we head east. Jay even buys a small guitar to learn how to play along the way.

We meet back up with Arthur from the train, along with his wife and kids, and stay a night with them. The next day, Jay's invited to a shanty town to view a suicide that Officer Arthur's working on. Happy to stay behind with the wife and two toddler daughters, I hope I never see a dead body. Ever.

When they return, Jay tells me Arthur had his gun loaded and cocked in his hand facing it out the window the entire time he drove around the shanty town.

He told Jay if on our journey we hit a black person, *just keep going*. Head to the nearest police station and tell them where it occurred. *Do not stop to help*, because there is always (*always?*) a group of black friends in the bushes waiting to jump you, rob you, rape you, kill you. *Ahhhh, something lovely to fall asleep to.*

I head for a shower, but instead find that his wife ran a hot bath for me in their miniature-sized tub. She asks me to save my dirty bathwater so they can bathe in it too. *Well, that's a new one for me.* The whole family shares the tub, one after the other. I feel a little selfish that they allow me the clean water first. Arthur explains the average water heater is just one tub full.

Grand goodbyes and we drive off.

Jay and I enjoy a few nights camping in a lush meadow with wild ponies, geese and chicken, stray cats and dogs. The smell of a controlled farm burns in the distance while I hang out the wet laundry to dry in the sunrise.

I meet a couple, Lilly and Luther, about 85 years old, who are camping too. They invite us over for some tea and egg salad sandwiches. They tell us all about their lives in Pretoria, about the apartheid, and that they would never want to come to America.

"Oh? Why not?" I ask gently, curious.

"Because you'll be driving down the road and get pulled over by a small-town police officer. He'll just break out your headlights with his nightstick and throw you in jail for the rest of your life!" Lilly stated as fact.

I smile at Jay, "That would never really happen," I tell her, assured she's been watching too many American movies. '*Footloose*' anyone?

Out of the blue, Lilly brings up rape and race. Again, she tells us her expert opinion, "Well it sure is a crime if a white girl gets raped, not so much if it's a black girl."

Wait, what?

Jay and I steal a glance at each other. "As a woman..." I sigh, struggling to find the right approach. "How... how can you even... say that...?" I ask slowly, sure the disbelief is evident on my face.

She goes on, with Luther nodding in agreement, "Well, a little black girl has been raped by her father, her uncles, brothers, neighbors already. She's had AIDS for years, so what is another rape? While a white girl who's brought up Christian, saving her virginity for marriage, helping her community, it's just devastating to her moral values. I'm sure you can plainly see that."

"Well..." I plaster on a slow-motion frozen smile, standing to leave. *I cannot share another breath of air with this woman.* "It's getting late. Thank you for the sandwiches and tea."

And with that, J.J. and I head back to our camp, stunned into silence.

I'm no detective, but... maybe, just maybe, the little black girl is fucking tired of getting raped.

Garden Route

- Sweet Jane -
Cowboy Junkies

Along the drive we spy Robben Island, home of the famous jail where Nelson Mandela served his 27 years of hard labor for a plot to blow up parliament in an attempt to end apartheid. In some of the bombings, women and children were killed. 19 people died with 217 wounded in the Church Street car bombing alone.

The gas station attendant tells us, "Back before Mandela got arrested, he used to put a burning tire around his enemies' necks, called it 'necklacing.' He loves Fidel Castro, Muammar Gaddafi, Yasser Arafat, he takes millions in donations from them and their terrorist organizations. Your American Oklahoma Federal Building just got bombed a couple of weeks ago, right? Try voting that McVeigh guy to be your President 27 years from now."

Wow. No one ever talks about that.

We'd been taught Nelson Mandela was like Abraham Lincoln.

We drive up the coast and through the Garden Route towns like Knysna, which boasts a sole wild elephant that roams the thick forest and is mysteriously seen only once every few years. *African version of Bigfoot?* Victoria Bay offers a little red steam train chugging along the hillside, protecting the tiny cove with a small row of vacation villas along a surf break. Plettenburg Bay is gorgeous with its peninsula views at sunrise.

Continuing east, we see giggling little African kids running barefoot behind a large truck that's hauling what looks like nothing but piles of sticks. The truck stops at the next stop sign, and each kid sneaks a stick out of the back as it's taking off. Giggling, they gnaw on the ends, and we realize it must be raw sugar cane.

Brightly dressed mothers work in the cane fields. They carry little babies and toddlers wrapped in blankets around their chests. Imagine you've wrapped a towel around your body after a shower, but you have a three-year-old on you, piggyback style. All day long they labor in the fields, working harder and lifting more weight than many of the men I know back home.

We meet some new friends as we unpack at a local hostel. Jaimee, a dark haired 28-year-old from Canada on her adventures hitchhiking around Africa, and Lexi, with sparkling blue eyes, big chipmunk teeth, and a crazy sense of humor. Lexi's originally from England but has been living for five years in Australia. I'm immediately in awe of girls traveling alone. We all decide to check out the Cango Caves an hour away the next day. Lexi insists I wake her up *any which way I can* so she doesn't miss out.

"Lexi, wakey-wakey," I try the next morning. She doesn't budge. I continue, "There's a huge breakfast buffet this morning. A nearby restaurant has donated free bacon and eggs, waffles and pancakes with fruit, and whipped cream. You've got to hurry, everyone's grabbing some!"

She leaps out of bed, still just in her undies and T-shirt, sprints to the kitchen to find an old, stained coffee pot filling up... and nothing else. Her face drops. "What about pancakes? Fruit?"

I grin. "You told me to wake you up any way I could!" She looks devastated.

"No bacon?" She rubs the sleep out of her eyes but jumps in the shower to join us on this drizzly day.

We drive down a dirt road through the Irish-emerald mountains with rocky cliffs down one side.

Lexi mentions she has one hit of acid, just enough for herself if we all don't mind. *Nope, go for it.*

Soon, we pass a "Ride an Ostrich" farm and Lexi squeals with laughter at the thought of all of us riding one. But Jay doesn't want to spend the money, plus it's raining pretty hard now. I regret that we skip it. How many times in your life do you get to ride an ostrich? The clouds are funny, my hair is funny, the fact the car only plays classical music is funny. Lexi's cracking us up.

The Cango Caves are amazing, but with cement stairs and concert lighting, it's pretty touristy. The others are bored quickly with the guided tour. They goof around as I go forward on my own through some cool cave spots. One tight spot is called the "Chimney" where, if I weighed 20 more pounds, I would've been too large. Two teenaged boys and I go in. I have to inchworm my way up 30 feet of a narrow rock formation to get out. It's hot, steamy, and claustrophobic. So, spelunking is now off my bucket list.

On the way back to the hostel, Lexi tells us her Nepal travel story, where she was on an old local bus, on the curvy road along the cliff's edge.

The bus slid out of control. Its back wheels hung over the edge. The front wheel wedged into a rock pile. As the bus slid even more vertically, Lexi was smashed against the very back window with all the luggage and people on top of her. Pinned underneath in a fetal position and she couldn't move.

Face to face with the 500-foot drop, whispering her prayers, she realized that window was the only thing holding her body inside the bus. The glass could no longer take the pressure and began to crack. She felt her body lose the stability of the shattering platform while suitcases slid into the abyss.

As she began her surreal fall from the bus, she felt strong hands grab her wrists and hoist her up with all their might. She was pulled clear from the vehicle and the entire bus fell over the cliff within seconds.

Lexi was safe along with the local passengers on the side of the road.

It sounds like a Hollywood blockbuster movie.

Jeffrey's Bay

- Gangster's Paradise -
Coolio

J.J.'s an avid surfer and he's invited me to paddle out a few times and share his boards. We agree Jeffrey's Bay is our next destination, and we end up at Van's Place Backpackers. It has five small bungalows, each with two bedrooms, a small kitchen and bath, with all the back doors facing each other in a small grassy courtyard. Home, sweet home for the next month.

We're just one block from a world-renowned surf destination, known for its perfect long right breaks, called "Super Tubes." I'd never heard of it before, but I learn to surf all the way to the sand just next to the angry reef.

We meet some great people staying there. Silas and Taran, both tall, shaggy brown-haired, smiley guys. Taffy, the pixie-faced girl with short brown

Tinkerbell hair who giggles contagiously. And her boyfriend of five years, Kaiden, who has long dark hair to his waist and looks like he sings for the Red Hot Chili Peppers. All in their early 20's. Fabiana, a quiet librarian with black hair wrapped in a bun, and her man Dale, with a receding hairline and bright blue eyes. Both in their mid-30's. They all hail from Australia. British 19-year-old boys, Noah and Grant, traveling with serious Ethan from Durban, and the peaceful, long haired, bearded hippie Byron, from New Zealand. We share travel stories and laughs, becoming a real community over the weeks we all live together.

One windy, huge-wave day, everyone heads out for a proper surf. Knowing I still suck, I hang back to write letters home. I'm calling home once a month, but Mom would prefer once a week. I figure more letters is a nice compromise. Plus, I'd accidentally showered with a very large, very scary tarantula the other day. I'd run from the outdoor shower, still wrapped in a towel, shampoo suds running down my face. I'd passed everyone in the courtyard with a scream, and it turned into a tarantula-throwing contest that I wanted no part of. I knew I'd be writing this story home. My own grandmother, Nanny, had a shower tarantula story from when she lived in Africa... it ended with Grandpa and a shotgun blast.

I sit at the dining table and begin writing, noticing the front door's unlocked and the back door is swung wide open for the breeze. Within minutes, there's a knock at the front door. The hair on the back of my neck edges up. For some reason, I hesitate to open the door... instead, gently pulling back the curtains at the window to the left of the door. Through the bars on the glass, I see two black African men standing on the porch. Before I can say a word, they both start shouting.

"My sister is bleeding! We need your car! We need your car! She is bleeding! Help us! Open the door! OPEN THE DOOR!" they both wave their arms in distress.

My first thought is *Your poor sister, what can I do to help?* Yet, out of my mouth comes Spanish. I speak *Span-glish* at best, but two phrases have stood out in my memory as an old joke.

"*Mecago en tu boca veinte veces...*" I surprise myself in a tone that sounds like "*I'm sorry, I don't speak English.*" The men stare at me.

"Open the door," one of them growls.

"*Melimpio mi culo con la cara de tu madre...*" again, as though I don't understand them. One turns to the bushes and slightly shrugs his shoulders as if to say, "*She doesn't understand, what should I do?*"

The bushes outside my new home shake violently as six other black men bolt from the scene. One of the men still on the porch looks at me, my eyes as big as saucers as it's dawning on me what's happening. Both men turn and run away, scattering. My left hand is still holding back the curtains, but, like lighting, my right hand flies up and slams the front door lock into place.

OHMYGOD. Where did the eight men run to? Do they know the back door is wide open?

I'm trembling as I run to the back and tackle that door shut, locking it faster than warp speed on the Millennium Falcon. I move chairs underneath the doorknobs as barricades and grab a huge cinder brick that had been a door stop. Seizing one of the many glass beer bottles from last night still lining the kitchen countertop, I firmly smash the butt of the bottle to make a weapon of sorts from the neck handle. I just sit at the table, alone for the next couple of hours.

Are the bad guys coming back? I'm so frightened, I don't know what to do. There's no phone. So, I sit, petrified, and listen for any sound, as though I could suddenly master Krav Maga.

I analyze exactly what just happened. These guys know this is a world-famous surf spot and when the waves go off there's always a surfy girl left behind. And today that surfy girl is me. If they really had a hurt relative, why go to the tourist hotel to ask for help from a stranger? Don't you have family or friends if you're a local? Isn't there a doctor in town somewhere? Call an ambulance if need be?

I did not even think. Spanish had flown out of my mouth effortlessly and the fact that I didn't speak English threw them off their illicit plans. I have a serious Guardian Angel looking over me. My naiveté and blindness to the world's problems are not to my detriment on this day.

If they'd even simply tried the front doorknob or just walked in through the wide-open back door, I'd have been gang raped by all eight men and Jay would've found my bloody, beaten, dead body

As I sit there with my brick and my broken bottle, it dawns on me what I'd told them, and I chuckle. Just a giggle at first, but then, belly rolling laughter with tears streaming down my cheeks as I realize what I'd said. The phrases were a hilarious joke at the time I'd learned them from a friend. They'd been forced into my brain, like I had no choice.

I'd said, "I shit in your mouth 20 times..." in my gentle tone, then, ever so sweetly, "I wipe my ass with your mother's face..."

Once the surfers return, everyone agrees I should not be left alone in the house again. Everyone's extra nice to me, but hippie Byron is especially kind. He gives me a huge hug and tells me he's so glad I'm okay. He hands me a note that if I should ever find myself in Australia or New Zealand to call on him or some of his friends, I'd always have a safe place to stay.

Spanish saved my life today. Of that, I'm sure.

*****

Sometimes I read my books on the beach while Jay's surfing, or swim laps in the man-made stone swimming pools that allow ocean waves to flow in. It's such a small surfing community that I keep seeing the same people over and over at different spots along the coast.

A brood of five blond, ripped surfers—Shane and his pals—had introduced themselves to me a few days back, and today I run into them at the market. They're hosting a barbecue, a *braai*, and would like it if I joined them. *Hell yes!* Taffy and Kaiden are leaving for Namibia, Jay feels like learning his guitar with Taran, while the others all shave Silas' head, so I go alone. I walk the few blocks down, looking over my shoulder for bad guys the entire time.

Shane and his friends say they're all pro-surfers, they get paid by sponsors to travel the world and surf. I must seem skeptical, because they all break out the surf magazines with themselves in the pictures. We play what seems to be hacky sack, but with a big volleyball, where we all stand in a circle and bounce the ball over our feet, elbows, even our heads, so it doesn't touch the ground. The other girls are friendly but don't participate. When the ball bounces off my head, they all call me a *bru*. Must be like a *bro* in California, or a *bra* in Hawaii. We all share laughs and great food. I'm happy I'm a *bru* at an African *braai*.

It's my first adventure after my scare, and it makes me feel a little more powerful to continue living my way.

Port Elizabeth

- Just a Girl -
No Doubt

It's Jay's birthday. To celebrate, we invite Dale and Fabiana from Perth to come along to Port Elizabeth to visit the Ado Elephant Park. This is why I'm here in Africa. The Wild. There won't be any lions at this park, but definitely giraffes and zebras, gazelle, kudu, hippos, water buffalo. We're all tickled pink to drive "Tina" on our own private safari.

It's a bright sunny day and we've packed enough water and snacks for a small army. We don't want to miss anything. Inside Ado, on the dusty dirt trail through the brush, we laugh as a group of eight ostriches do a slow jog along the road in front of our car. They seem nervous that we're following slowly behind them. They deliberately keep glancing back to see if we're still there, but simply refuse to go into the bush for 15 minutes.

For a full hour we play the game of *we-almost-think-we-see-an-elephant*, with a tree branch bent like a trunk, or a huge boulder looks like it has legs. It's getting dull.

Dale shouts first, "Just there!"

Our car collectively holds its breath. We peek through a group of small trees, finally seeing our first group of elephants. No matter how many zoos you've been to in your life, no matter what you think you'll experience when you first see large animals in the wild, *it's pretty damn exciting.* All our hearts beat loudly in our chests, our faces flush, all four of us giggling like teenagers on a first date. There's a mama scrutinizing us as her baby runs just two feet in front of our car. Fabiana and I share a teary-eyed smile.

There are 19 elephants traveling together as a large family. The herd keeps a slight distance from us. We can tell these magnificent creatures are wary, especially with a newborn. The group instinctively circles the baby and begins moving away from us. Peering from the car windows, we all notice that a few of the elephants seem agitated with one particular peer. They keep turning back toward him and trying to push him away. They wrap their trunks around his trunk, getting up on hind legs to shove him backwards. He just hangs back for a moment, then sheepishly follows behind.

They're walking pretty far off the road now, and we're not done with them yet, so Jay and I get out of the car and tentatively follow along. Dale and Fabiana hang back to make out in the car. Hey, no Big Cats here after all, what else could possibly go wrong?

We're now quite close, within 20 feet. We keep snapping close-up pictures of the wild beasts in their natural habitat. Some are still pushing the one away, but he keeps persisting. Their slow steps are uncharacteristically silent. Their huge, flat feet muffle the sounds of sticks breaking. *This is a Gift from God just for us.* The goliaths don't seem particularly bothered by us. We must not be the first to climb out of the car to join them on their walk-a-bout. *Like your trip, Mom?*

The herd wanders much farther off, now just a small row of dots in the distance. We look back at our car and she's just a dot of red in the opposite direction. Man, we've followed the elephants quite a fair way. This area has no big trees or big boulders, just smaller and medium bushes scattered for miles along the arid outback. Jay and I start goofing around by throwing small rocks at each other as we mosey back to our car, cracking jokes, laughing.

The ground begins to shake. Just a little at first, and both coming from California, we assume just a small and harmless rolling earthquake? The shuddering gets stronger, until we both get the same idea and turn back toward where the elephant herd had disappeared from sight. Now they're a dust cloud. *Heading. Our. Way.*

They're chasing that one in a beeline toward us. *And they're fast.*

"RUN!" Jay shouts.

The gray giants are getting closer and closer, right on our heels. There's a thunderous rumble accompanying the out-of-control stampede. Loud trumpeting. The flurry of soil they're kicking up could match the Baja 1000 car race in Mexico. Bushes are trodden like a pile of toothpicks thrown into the air. I'm running faster than I ever thought I could, jumping over bushes and swallowing dust.

I *can* run, but unless it's to the phone or the ice cream man, I choose *not* to run. Yet, until faced with something to help you decide, you never know if you're a man or a mouse. *I sure could use a mouse right now to scare the elephants.* Adrenaline kicks in.

They're breathing down our necks. Their dust cloud encompasses us. We're still 15 steps from the car.

There's nothing we can do about it.

They're here.

The entire group of furious, stampeding elephants veer left, just behind Tina. Jay and I are awarded the privilege of our last few feet. Veering to the right, we jump into the car with the steamed-up windows, Fabiana and Dale and the boink be damned.

As we drive off, flustered, filthy and in awe, explaining our adventures to our pals, Jay makes a quick stop next to a small coil he spies on the road. Oh, just a Puff Adder snake, one of the most venomous in the world, that's all. Good thing we didn't jump onto one tearing through bushes, moments ago. Quick picture and we speed off.

Almost trampled by wild elephants on your 22nd birthday? Best birthday ever.

The Transkei

- Flight of the Bumblebee -
Rimsky-Korsakov

To get to Natal in the east, we're warned that instead of driving five hours through the Transkei region, we should drive 25 hours to go completely around it. It's gorgeous, but it's that violent. *Hmmm, 25 hours, though?* We decide to rush through the dangerous, yet beautiful Transkei. We'll go as quickly as we can, not stopping for any reason, except fuel.

It looks exactly the same as the rest of South Africa. Picturesque, yes, but not extraordinarily so. We pass old burned out car skeletons, I'm sure with unspeakable stories to tell. But nothing happens to us, so it doesn't seem that dangerous either. I help Jay practice his Indonesian phrases to pass the time.

I'm nauseated thinking about white rice again, so I insist on stopping at the first market we see once we're through the Transkei. Delighted, we find little powdered soup packets we can call gravy for the rice.

I spontaneously opt for a tiny bottle of red wine with some cheese and crackers, and even Jay jumps at the chance. We savor it on the cliffs overlooking the waves, listening to the car's classical melodies wafting into the sea breeze.

The news crackles through the radio speakers. Five British travelers had car problems and broke down in the Transkei just yesterday morning. All five men and women were raped and murdered, burned alive.

The wine sours in my throat. Jay and I share a teary look as we wonder if it was any of the cars we'd passed. How are we still here? Do we just keep going on? Enjoying our lives, until it's our turn?

Absolutely tragic. I am truly humbled yet again.

Durban

- Badfish -
Sublime

Passing through Amanzimtoti, we hit a few surf spots before ending up in Durban. It's much warmer with a tropical feel, skyscrapers line the beach, comparable to Honolulu or Miami.

Drumbeats fill the night air as we walk through the city that night. We stumble upon 25 Zulu tribal dancers in full costume performing in the streets, while tourists gather to throw cash into their bucket. When my Nanny and Grandpa lived in Durban, maybe they stumbled across authentic dancers in the 1930's.

We meet friends who open their homes to us. One gentleman is a white-haired ship's captain from the South African Navy. A man of maybe 75 with his teenaged son. We're invited to a musical jam session at someone's home and we encourage them to join us.

The jam session is—it turns out—at the home of professional guitarist Mike Rutherford from the band *Genesis*. There are only 30 of us, so we get to chit-chat with him and get a personal concert. He plays a lovely version of The Beatles' "*Here Comes the Sun.*" Someone else sings "*Wooly Bully*" and Jay goes nuts.

"They don't know!" he shouts. "They just don't get California like we do!" *Wow, it's true...* we're sneaking a glimpse of their culture, but they haven't felt ours.

A few people are smoking joints, being mellow, while others take turns playing guitar and singing. I'm drinking a glass of red wine. The Captain is the only one who is visibly drunk on hard liquor and takes it to the next level. To his son's dismay, he's downright annoying and I'm embarrassed we brought him.

Driving back to his house, he's swerving, shouting, "I was a Captain in the Navy! I have commanded many ships! You must call me Commander! You WILL salute me!" His son squeezes my arm in the back seat, shouting back, "Yes Sir, Commander, Sir!" The young son had already asked his father not to drink at the party and, clearly, this happens a lot. Poor kid.

Jay puts up with this for exactly 30 seconds before he kicks him out and takes over the wheel, all while complying with the "Yes, Commander Sir!" bullshit. Still shouting and slurring from the front passenger seat, his arm up in a salute, the pathetic former captain quickly passes out. We move out in the morning.

Then there's Shannon, an older surfer lady with a tiny frame and long sandy hair. She runs a souvenir shop on the side of the road. She offers us a place in her home with her very quiet, downright creepy, husband who doesn't say one word to us. He just stares.

She takes us out back and introduces us to her enormous dog, a Rhodesian Ridgeback with a huge wound down his thigh from recently fighting a wild boar. Shannon says not to worry about her husband... he's just super shy... and high. She tells us that smoking pot in South Africa used to lead to a mandatory three-year sentence in the insane asylum. She was sentenced years ago and had to fight off all the lesbians who wanted to "get together."

J.J. and Shannon really bond quickly over the waves. She's been an avid surfer too. She invites us to see the movie '*Woodstock*' with her later that night. After the movie, back at her place, Shannon stays out to feed some animals while Jay and I head in, but it's pitch-black and we're stumbling over furniture. Her dog lets out a deep growl, and he's barking. The hair on my neck stands up. In the darkness, we can hear him but can't see him. He keeps falling because of the injured leg. He's trying to snap at us while Jay and I keep huddling together, slowly backing toward the door, not sure if we'll be bitten any second. *Yikes*. Shannon comes in and handles it, but it's time to move on again.

Through Shannon, we meet Vanessa and her three daughters, who invite us to stay. In her mid-30's with short brown hair, bright blue eyes and a contagious laugh, Vanessa was widowed when her husband died in a motorcycle accident. Her little youngest blondie daughters, Evie and Charli (maybe four and five years old) sadly won't remember much of their father. Poor darlings. But the love in this family is felt for miles in each direction. Her oldest daughter, Camryn, and I truly hit it off. She's 17 with long blonde hair, a megawatt smile, and her mama's eyes.

And she surfs. We eagerly hit the beach in the morning. She's so impressed with Jay's surfing that she squeals in delight, "Oh my god! He just did a 360 Floater! He's amazing! No one does those!"

"What's that mean?" I ask, lazing in the sun. She explains it's where after surfing in the middle of the wave like normal, he jumps his board to the top, right where the wave is cresting over, does a complete spin with his board, and drops back into the middle as if nothing happened. He's been doing these all the time and I had no idea how skilled he was.

Vanessa's learning the guitar, so she and Jay spend the afternoon practicing. That night, she takes us to see her kids at a school play. It feels so natural to be included as though we're locals... the little girls even fall asleep on us in the back seat of V's car. We share a glance and giggle. I ponder how my life had changed so quickly from living in a ski-resort town for a season, to be so lucky to have all these new adventures under my belt, and now I'm holding a sleepyhead darling angel in my arms on the way home from a kids' school play... *in Africa*. What a whirlwind.

With an agreement to come back, we move on to stay with our moody pal Ethan, whom we'd met in Jeffrey's Bay a few weeks back. Mid-30's, tall and fit, with light brown hair yet a serious outlook on life, he offers a couch for us to crash on. His place is modern, in a gated community, a white condo with dark brown wooden beams everywhere. Instead of showing any interest in us, upon arrival he plugs in his electric guitar and practices loudly for hours. We just have to stare straight ahead and not speak, I guess.

He finally takes us out on the town that night. At a bar, the crowd has an equal men-to-women ratio until exactly 9:00 p.m., when all the women suddenly clear out. There are literally two of us left, along with lots of professional Springbok Rugby players.

After playing pool with Jay and Ethan, I head to the ladies' room, only to get cornered by a huge, muscly, rugby player. He's very flirty, but aggressively so. His arms are blocking my way and I'm pinned against the wall like he's going to kiss me any second.

Over his shoulder, I see Ethan is fully disgusted with this scenario, and not enjoying the Springbok's belligerence myself, I break free and go back over to my group. Ethan's now very harsh with me and tries to leave me at the bar. He keeps saying I should catch a ride home with this rugby player.

I'm confused.

We head to a different bar, only now there's tension between Ethan and me. We play pool and drink a bit more, but I'm ready to call it a night when Ethan runs into a pal of his. Stanley, a little guy in a suit, seems nice. Jay's still having fun playing pool, so Stanley offers to drive me back to the house, and since Ethan's quite cold to me now, I agree. It's been raining quite heavily tonight, so Stan pulls his car up and I climb in. I'm not a car person, but this is a black, flashy, new car, like a Porsche. We immediately get on the freeway, yet... I'm sure we didn't take the freeway here.

I realize too late that Stanley's wasted, slurring his words, swerving the car. He puts his hand on my thigh instead of the gear shifter. I SCREAM in his face to pull over. He does so in the emergency lane on the freeway in the pouring rain. Déjà vu from Jay and the Captain, I get into the driver's seat and shove him across... he's like a wet noodle. I'm now driving the car on the opposite side of the road from what I'm used to, with a stick shift also on the wrong side, so the numbers are reversed in my reverse hand, on the freeway in a foreign country—oh yes—and in the rain with a drunk, horny stranger. *Well done, Pam. Bravo.*

I don't know where Ethan lives and I'm just looking for anything familiar when Stanley slurs, "Exit now!" I swerve across three lanes and exit. Stan directs me a few streets down to the gates of Ethan's condo.

"Oh, forgot... gate code...," he murmurs, half asleep. I park sideways in front of the gate, jump out of the car, leaving the door wide open, and in the beam of the headlights, I climb a tree near the fence. The rain's splashing into my eyes, mascara running down my cheeks, my hair slicks down. My foot slips, but, just in time, I strap my leg over the top of the barbed wire and slide down the metal fence like a fire pole. Fuck you, Stan the man. I run to Ethan's place a few buildings down and the light's already on. They're back as well. I stumble through the door looking like a wet rat. A *furious* wet rat.

The guys are looking at me like *what the hell happened to you*. I explain.

"Well, all the nice girls leave the bar promptly at 9:00 p.m. and go straight home, while only hookers and sluts stay out late," Ethan arrogantly enlightens me. I guess I can figure out why that rugby player and Stanley were so interested.

"I was out late because I was with *YOU and Jay*. I don't know your rules! I'm not from this country," I grit my teeth. *Thanks a lot, Ethan.*

He gets me some Band-Aids for my legs, scraped on the fence. And a towel. *Dick.*

I'm ready to move on, but Ethan wants to make it up to me and drive us up the coast for a few days to Sodwana, a beautiful stretch with no tourists and great surfing. Jay's excited to go, so I grudgingly accept.

We sight-see, stopping on the true spot atop the cliffs that the famous Shaka Zulu warrior king—upset from the death of his mother—slaughtered thousands of his own people in the early 1800's. It's surreal to stand on the bluffs overlooking the rocks where real human beings— men, women, children, and even their cows— had been forced over the edge to fall to their deaths. I may have earned straight C+'s at school, but this history lesson I'll never forget.

Camping under the stars, sleeping in the back of Ethan's open bed truck is okay. The waves are too big for me, so the guys surf each day, while I read my books and write letters. We see no one else on the beach for two days. Jay and Ethan practice guitar, while I cook the rice. Ethan hates the flavor, which secretly pleases me. I help Jay with his Indonesian language phrases. It's beautiful, but the weird vibes with mood-swing-Ethan make it less than great. When it's time to drop us off, he peels out in a grumpy mood again while Jay and I wave goodbye.

We turn to each other and laugh.

In Durban, we hang out with Camryn every day. I tell her about the 9 o'clock thing I just learned, and we laugh about the subtle differences between our countries. I explain how in California, if you walk down the street and pass a handsome stranger, after a few steps, you look back and check out their *assets*. And you hope they're doing the "Look Back" to check out yours, too. I'd just assumed this was human nature, but alas, in Africa every time I do a look-back at someone, they aren't looking back at me. I'm not feeling very attractive, but Cammie and her friends have never heard of this tradition.

I tell her once, at a barbecue, I asked to make a local phone call and after I hung up, I got a few dirty looks. Not sure what I've done wrong, I ask Camryn about it. She mentions there's a coin jar by each telephone for me to put money in. I'm happy to oblige but tell her I'm just used to paying a monthly flat fee for as many calls as you want back home in the USA.

And "Let's go *just*-now," means I'm the loner who keeps jumping up. It means everyone sits around another hour before the final shout "Let's go *now*-now," because that's when people are *really* ready to go.

*****

"My sister's getting married!" Jay shouts enthusiastically after a call to his parents.

"How cool!" I reply.

"Yeah, he's a great guy. She's getting married soon. Like *soon*-soon. I think if I leave South Africa next week, I can still get a couple of weeks in Indonesia before the wedding..." Jay lets that last bit dangle out there as it takes a few minutes to sink in.

"Oh... you're going early? I don't think I'm ready to go home yet!" I panic.

"No way, you don't have to leave just because I'm going. Lots of people travel alone. You've got friends here and you've learned a lot. You'll do great," Jay reassures me.

"I just have to process this. I wasn't planning on traveling alone through Africa!" I'm annoyed. I'm forced into an uncomfortable situation. As if I'd hired him to be my personal tour guide.

"You're going to have FUN!" he says, with a laugh, while I scowl.

Just our luck, a friend of Vanessa's in Jo'burg wants to buy our car for nearly the same price we bought her for. Easy Peasy. Jay can drive and meet him near the airport.

So, with a hug and a few tears from me, Jay's off to his next adventure and I'm on to mine.

As I watch my dear friend wave out the window with his blond ponytail blowing in the wind and his surfboards hanging out the back, it hits me that no one on Earth will ever know this South African adventure side of me better than Jay. We shared something rare and precious. He taught me how to eat for pennies, launder my clothes with a bar of soap in the sink, how to open a can of food with just a knife, how to be vulnerable and respectful of new cultures, how to think about safety and how to see the world with childlike wonder.

After helping him practice his Indonesian phrases the whole trip, the perfect one comes to mind as I blow a kiss into the wind and whisper, "Thank you for everything Jay, *sampai kita jumpa lagi!*"

Translated perfectly for that moment to "*Until we meet again!*"

After six weeks together, I am now officially—deep breath—*traveling solo.*

*****

The Gunston 500 surf competition's coming to Durban. I've been invited to stay for a few weeks at Camryn's friend Heath's house, right on the beach, overlooking Joe Kool's Bar and Battery Beach pier. A bachelor surf pad with a mattress on the floor in the sunroom for me. Camryn must finish up at home and can join in a few days.

"Just crawl through the kitchen window if you forget the house key." Heath is a gorgeous 30-year-old JFK Jr. lookalike. Yep, not a bad spot at all... *except* for his flatmate. He's a very short guy with an obvious Napoleon complex. He keeps bragging about how many drugs he's on at once. "I did so much blow last night, but I had to get some sleep so I took a handful of downers but that made me feel strange so I had to balance it out this morning with three Ecstasy pills..." *Yuck.*

The bachelor pad could use a deep cleaning, and since I'm their guest, I think I'll do them the favor. They're having a braai this afternoon, so after everyone's done eating, I clear all the plates into the kitchen. As I begin the dishes, Druggie Roomie yells out to me between puffs. "You know we have a native maid for that..." he smirks, arrogantly.

He proceeds to tell all of us, "All black people should be white's servants, they're all just animals. When South Africa was discovered by the Dutch, they hadn't even invented the *wheel* yet. And here we were, wearing clothing already, sailing around the world on ships! Look at the shape of their skulls, it's different than ours, their bodies too, the shape of their noses, even their hair is like fur. They're not even human..." His friends look at each other uncomfortably, but *no one says a word.*

My blood boils. My face turns red. My heart beats in my throat. I put down the sponge, open the front door and walk down the street to a crowded coffee place. My chin quivers as the tears burn silently into my cup of tea. I can't believe how ugly this one useless speck of flesh can be.

Someone clears their throat. I look up into soft blue eyes, long curly wheat hair and a warm smile. A girl asks if she can join me. I nod and smile through my tears.

"Hi there, I'm Rebecca..." she sips her tea, a look of concern on her face. After I introduce myself, she inquires as to why I'm upset. I tell her the entire horrific conversation.

"And the most upsetting thing is that no one said *anything* to him to shut his mouth or to counter his crappy details. I wish I had the words..." I say.

"Well, I'm a Red Cross humanitarian aid worker from England, and I've studied Anthropology with specifics to countries in Africa. My mother came here as an aid in the 70's. I can tell you that their skulls are shaped differently, smaller, tighter, so their brains won't boil under the African sun. That's why their hair is different as well, to keep its shape even in the rain. Large predators are around every turn, so their muscles and bodies have evolved differently so they can run faster, plus historically they must walk many miles looking for water. Their noses are wider, flatter, because warm air naturally rises, but in England we have pointed noses to try to suck cold air up into our nostrils. We *had* to create clothing, build homes, invent things or we'd freeze to death in the snow in Europe. They had plenty of food they picked from the trees and bushes. They're traditionally hunters and gatherers, so they just needed some shady huts, and a loin cloth or so. Over thousands of years that didn't need to change."

I can't believe my luck to have someone so knowledgeable and scientific sitting next to me at the exact moment I need to hear this.

She continues, "They were happy here until they were 'discovered.' Then here we come telling them they should be ashamed of their bodies, to cover up. That they're farming wrong and they should do it our way, that they need our medicines when they had it all in their indigenous plants. Do you know companies like Nestlé gave them free infant formula by the millions, telling the native women how much better it was for their babies? The women began to use it, allowing their breast milk to dry up. Then the company said, 'it's not free anymore' and forced these poor families to shell out money for food that was God-given! Isn't that despicable?"

I buy us both a muffin and ask her to tell me more. "They're a very proud people. Have you noticed they speak very loudly? They'll yell across the street to each other. They believe it's very rude to talk softly, they think it means you're whispering negative things about them, so to prove they aren't, they project their voices to show transparency. And they *have* invented many things. The elevator, car gear shift, furnace, clothes dryer, mailbox, blood banks, the pacemaker, even a clock... just to name a few."

I laugh, "Where were you 20 minutes ago, when I needed you!"

She tells me it's not all roses though. Her mother stayed for two years and couldn't do the help she'd wanted, mostly because of the indigenous attitudes now. She tells me as she begins her own two-year commitment things haven't changed. Her job is to see a tribal town might have extra bunny rabbits around, and remind the elders how their ancestors used to make bunny soup, but the tradition has not been handed down due to the western intervention. Her team will teach how they used to farm with wheat or vegetable seeds, harvesting half for food and re-growing more with the seeds from the other half. Then, the Red Cross would move on to another town, see their resources and remind the locals of their traditions.

"After about six months, the team would circle back to the first village to see how it's going, but no one's making bunny soup. And they'd have mowed down the *entire* harvest, not replanting. They'd hold out their hands and demand more seeds again. So, the Red Cross would show them *again* all the details and *again* circle back in six months and still, no one would be doing it, they would demand more seeds. That can be tough, to not get through, when it would work if anyone would just take the lead in the village," she shrugs and sips.

She continues, "Many of the villagers have HIV, or full-blown AIDS. They somehow learn from their tribal doctors that if they have sex with a virgin, it'll cure their diseases, so rape is prevalent. And victims are getting younger and younger, to make sure they're getting a virgin, they'll even rape a baby. They don't realize they themselves already have AIDS and have just spread it to the victims."

My stomach turns, I'm sickened. "A baby? But how? I mean... how can it even... *fit*?"

Her brows furrow. She swallows hard, leaning in toward me. "It doesn't. It tears open and creates a cavern that needs surgical repair. It's called fistula. We help with that as well. I've seen it, it's horrific. Their culture is just so very different. The men don't like a lubricated woman, they think it's from whoring around, so to prove they're 'faithful' and 'good' wives, the women literally put a couple of fingerfuls of dirt from the ground up inside themselves, so it's dry when their husband gets home. The friction creates lacerations in both partners, and with open wounds, diseases are spread even faster than ever."

I'm in awe of someone making such a commitment to help the world while I've just been observing my way through it. She must go, but we hug, and I tell her I'll never forget her, she's opened my eyes about the true nature of things. I make a small donation to her cause.

What a heavy day. I head back to Heath's before it gets dark. He's there but everyone else has gone. Honestly, I'm relieved. I tell Heath about his flatmate's opinions.

"Oh, don't worry about him, he used to date a black girl, he was so in love. She dumped him for a black boyfriend, so he says he hates them all now, but secretly, he still tries to date them."

Well, that is an unexpected twist. It turns out the rude dude has gone camping to get away from the crowds during the *Gunston*, so I won't have to see him again. Yippie.

The world-renowned surf competition *The Gunston 500* arrives at our pier. Heath has a booth, selling T-shirts, women's clothes, and Smart Drinks of his own invention: flavored vitamin water, way ahead of its time. He's asked Camryn and me to help out. Yay, my first job overseas.

Everyone's out to make a few bucks with all the international tourists. For the entire next week, I'm on the boardwalk in Durban, enticing folks meandering by. The black locals wear their tribal gear and offer pull-carriage rides, or sell necklaces made of rolled up pieces of paper to look like beads. There are hand painted T-shirts and hand carved drums.

We're selling some awesome tie-dyed skirts from Heath's friend and we must keep calling her to replenish the supplies. She's bringing as many as she makes as fast as she can make them. Girls are literally fist-fighting over these skirts. What a little money maker she's got here. Hmmmm, maybe I can make skirts like this when I get home... *Ca-Ching.*

"You sure can tell you're American!" Camryn giggles.

"How so?" I was feeling like a local with a job and everything.

Copying me, she shouts in her best American accent "Hi guys!" to people walking by, selling them a couple of drinks. It turns out they don't say "guys" at all. Oops.

Camryn and I take a break from selling to go lie on the beach and watch the competition. I'm lying on my back but looking toward her with my arms crossed behind my head. We're talking about something profound, I'm sure, when she gets a case of the giggles. I'm watching her laugh and laugh, and I just don't understand where this is coming from since we were just talking about something serious.

Finally, she points and says "Look!" I turn my head to see a TV camera in my face. They've done a slow-motion film up the entire length of my bikini body as I was turned away. Now that they're caught in the act, I just smile and blow a kiss as they move on to more of the crowded beach. Now I have to laugh as I tell Cammie the truth.

"I haven't shaved my underarms in a couple of days! They must have snapped quite a stubbly close-up from that angle with my arms behind my head!" We burst out laughing.

"I'm never going to be shy again! Your reaction was perfection, Pammy!" she hugs me.

The Gunston 500 is a great success. I'm right in front of the giant orange stage at Battery Beach in Durban as Shane Powell is crowned the winner, but Kelly Slater, Timmy Curran, and other big names are celebrating their heats and spraying champagne on each other, and on us. I run into all my friends from the braai in Jeffrey's Bay, where we bounced that ball off our heads. Big hugs all around. And here I was worried that after Jay left, I wouldn't feel comfortable so far from home, but I'm having the time of my life.

Cammie and I go out to Joe Kool's each night to meet the surfers, play pool, and dance the night away. Heath also plays in an African band with indigenous guys from Swahili and Xhosa (pronounced *Ko-za*, the tribes talk by adding clicking sounds). We're invited to see the show and, man, are they cool. I can't pronounce the band's name, but it means "The Devil" in one of the local languages. Kind of reggae mixed with hard rock.

I forget Camryn's only 17 since she's allowed in the bars and no one's questioned her. It reminds me how backward our American system can be. You can get a job and drive a car at 16, but you can't see some boobies in an "R" rated movie until you're 17. And you can join the military and fight for freedom around the world, get married and have kids at 18, but in no way are you responsible enough to have a beer after work until you're 21.

*****

Spending so much time on the beach, from sunup to sundown, I notice the sun rises to mid sky, does a small circle, and sets very close to where it had risen from. I'd heard it circles the sky in the North and South Poles, but to see something close myself is quite intriguing.

Sitting on the sand, I reminisce on my family history. There's many moments wondering if my Nanny had been sitting on this same exact spot, watching

Grandpa splash around in these same waves. Grandpa had swept her off her feet. "Marry me and I'll show you the world!" he'd said to her. They married *three days* later and were off. Does love like that even exist anymore?

They're British, living first in South Africa, then in Egypt. While in Durban, the native women had been very fond of Nanny's clothing, so she donated dozens of house dresses to the ladies. But Nanny had an ulterior motive. This was a very conservative time in the 1930's, and coming from England, she wanted to get the ladies to cover their bodies, their exposed breasts. In town, she kept running into ladies walking with their children, and all were wearing her dresses. They ran up to thank her, feeling very fancy. But Nanny was still blushing. Each and every woman had cut out large circles on the chests of their new dresses so that their bare breasts still hung out.

As I pass indigenous families in the streets of Durban, I wonder if their grandmothers wore the dresses of mine.

If only I had a time machine.

In Egypt, my grandpa worked for Imperial Airways, where his job was to clear the hippos from the Nile River so the water planes could land safely. *How do you move a hippo, exactly?* He'd been hired as a tour guide for the real Lawrence of Arabia along his travels. They'd spent a great deal of time together as my Grandpa Bert showed him around for weeks. They were traveling with several dozen men who heard rumors that Lawrence of Arabia was in the country, not realizing that *Lawrence was with their group.* He was snoring in the tent behind them at that very moment.

My Auntie Anita, my mom's older sister, was born in Egypt in the 1930's on the Princess Ferial's first birthday. In my grandparent's home, my aunt was delivered out on the large veranda lounge with long, sheer curtains blowing in the African breeze. During her labor, Nanny watched as the celebratory fireworks exploded across the Nile River, above the palace, for the baby Princess's birthday party.

My Auntie's middle name is Ferial after the Princess herself. As a young child, Anita was invited to play with the Princess at the palace and they became quite good friends. My Auntie even recalled a few parties where a young boy of 12 held out his hand to dance with her. That boy was the late actor, Omar Sharif.

There's even a documentary film called '*The Last African Flying Boat*' that has some footage of Nanny and her three-year-old daughter, my Aunt Anita, disembarking from a water plane and dressed to the nines. What a glamorous time to be abroad.

My mom, Wendy, traveled to Kenya and Tanzania in the 1970's with her friend Mike. They had a few hilarious adventures too.

Mike set up camp and since they hadn't seen another person for days, when my mom happened upon an empty beach, she stripped down for an African skinny dip in broad daylight. After a peaceful swim out into the ocean, she dropped down on her towel and fell asleep. She was awakened by the entire troop of the British Navy, their boats pulled onto the sand, practicing their marches for the rest of the afternoon.

My shy, modest mother had to grab her clothes and slink off into the sunset.

I'll just bet the British Navy had a surge in recruiting new troops to serve in Kenya after they heard stories of a beautiful, blue-eyed, raven-haired, naked lady in the area.

Maybe she made the newest brochures?

One night, she went to collect firewood for their camp and her foot got caught in a hunter's animal trap. It swung her right off her feet and there she hung, upside down for half an hour as she called out for help over and over again, the blood rushing to her head.

Finally, when she hadn't come back to camp, Mike went looking for her and laughed another 10 minutes until she begged him to help her down.

Camping on the beach under the stars, my mother woke to find hundreds upon hundreds of dinner-plate-sized crabs crawling all over the beach, including on top of their sleeping bags.

She slept in the car after that.

In the city, the two had secured a hotel room and upon entering, my mom noticed a small button above the bed.

"What's this button for?" she asked as she pushed it a few times.

"Don't push that button!" Mike tried to catch her, but it was too late.

There was a knock at their door, and a young African woman stood with a shocked look on her face.

Turns out the button summoned the local brothel to send over a hooker. She was not expecting to see a white woman's face at the door, that's for sure.

*****

Camryn and Heath have also touched my life as they invite me to surf with them every day.

I feel like a surfer again, walking down the beach in a borrowed wetsuit with a borrowed board under my arm. Instead of paddling out and getting my arms all tired, I follow the Durban tradition of walking to the end of the pier, climbing over the railing, holding my board and my breath, and just stepping off the edge.

The board pops me back up like a champagne cork so I can catch the next wave in and run down the pier to keep jumping off all day long. Yep, jump off the pier to get some adrenaline before I surf in.

When the events come to an end, Cammie has to go back home to Vanessa's. We all agree that if I'm to see any more of Africa, I should go to a hostel and meet more travelers to join with.

Camryn drops me off with the promise of another visit before I leave the country in a few weeks.

Hluhluwe

- Hits from the Bong -
Cypress Hill

My first night back at a hostel I meet Tristan from Scotland. Tall, skinny with lots of fluffy blond curly hair. He hasn't traveled before either and has just arrived in Africa. He introduces me to James from England and a guy we nickname "New Zealand." They're renting a car tomorrow and going to see the Big Five Animals: rhino, buffalo, elephant, leopard, and lion, in the Hluhluwe National Park if I want to come. Yes! Yippie, the Wild.

Tristan finishes packing the rental car and off we go to see some African animals. James looks like a blond 28-year-old but carries himself like a wise old professor. I picture him at home in his full library wearing a sweater vest, smoking a tobacco pipe, and sitting in a wingback chair. I guarantee I'm wrong, but that's funny.

Meanwhile, New Zealand has bright green eyes and dark hair, and has traveled the world over his 30 years. He's quick-witted and quick to laugh. He wears a necklace that looks like a cross made from a piece of elephant tusk ivory, but he explains that it's a hippopotamus tooth. I swear it's the same size as my forearm. They've all just met as well, so we're driving and swapping life stories. Looks like I already have some new friends.

After driving out of Durban a few hours, we arrive at the entrance at Hluhluwe Park (pronounced *Shoe-Shlew-wee*) and, to our disappointment, it looks closed. We're confused, but realize from our map that we've just arrived at the back side of the park. It looks like a couple of hours to get to the right entrance.

As we're pulling the car around, we see a lone young black African man leaning against the fence, wearing that *'Dr. Huxtable'* sweater in 100-degree heat. New Zealand pulls out a plastic baggie of nothing but seeds. Weed seeds. He shakes it at the young Zulu man as if to ask without sharing a language "*Do you know where we can buy some?*" Zulu nods his head and ends up jumping into the car next to me. Again with the worry *is this a good idea*? Are these guys going to protect me from danger if I just met them *yesterday*? Are we *positive* the three-year insane asylum sentence is no longer mandatory for getting high? Zulu smiles a warm smile at me, so I smile back. We bond.

So, Zulu points us in the right direction and off we go, no longer in search of lions but the ever-elusive marijuana leaf. We're off-roading for another hour. He keeps pointing up a hill with literally no road and we all keep exiting the car to physically push it up over big rocks. We climb back in and off we go again.

Soon, we see an African couple, both wearing tribal grass skirts, no shirts, and dried white clay on their faces, necks, and torsos. The woman is carrying a humongous pot of water on top of her head. He's carrying nothing. *Husband of the year*? Frowning, they spot our car, holding their hands up for us to stop.

They are Zulus and we are trespassing.

Their body language is telling us to *Go back the way you came! How dare you drive on our land!* Randomly, they notice their friend in the car with us, and their body language goes instantly into a relaxed *Hey, no way, what are you doing with them? Why are you bringing them here?* They're laughing as our Zulu buddy explains that we're looking for some Zulu bud.

I notice the man leaning into the car window asking the questions but I'm hearing the *wife tell him what to ask*, just like a slightly annoyed wife would do in America. She'd say something then he'd repeat it, and our friend would answer. She wouldn't wait to see if her husband would think to ask another question, she would shout out her own question to make her husband ask it next. She definitely wears the tribal skirt in this marriage.

We wave goodbye, and soon arrive into a town of 15 large, circular, thatched-roof mud huts. We climb out to stretch our legs and see the sights. There are goats with their kids everywhere, chickens with their chicks wandering together, along with many, many dogs and puppies, cats and kittens. *The Wild?*

Obviously, everyone comes out of their homes to see why there are four white people with their friend. In their village. Everyone quiets down as the largest black woman I've ever seen in my life comes out of the nearest hut. She's dressed straight out of '*Gone with the Wind*' including the bandana wrapped around her hair. We're at Big Mama's hut.

Our Zulu friend doesn't say one word. He grabs the baggie of seeds from New Zealand and shakes it just once at Mama. She doesn't speak either, and listlessly goes inside.

Mama comes out, sluggishly dragging a full-sized-up-to-your-waist barrel of already pruned ready-to-go Zulu Ganja. She walks it back and forth until it sits right under our noses. It's full. And it smells *good.*

New Zealand arrogantly whispers in my ear, "Watch me bargain with her!"

Big Mama reaches into the barrel, elbows deep, and grabs a double armload. She holds it up in her apron and says in perfect English, "Six bucks."

New Zealand quickly shouts in a very excited little squeal "Okay!" like a middle school student who has found his first Playboy magazine. He peels off $6 and we grab a pillowcase to put it all in. Mama's son asks if we would like to stay for dinner, and I regret that we say no. That would have been quite a lovely and culturally awake moment away from my LA lifestyle to have dinner deep within this dusty African village, at the home of these fine people.

However, now we have some pot and my pals want to get high.

We decide to spend the night in a treehouse for an expensive $25 each for the night. I just go for it. What a great choice. We're inside the correct entrance of the park. We can wake up in the morning, walk around after breakfast tomorrow, and drive our rental car through our own safari.

A bamboo ladder leads us up to a small platform patio connecting two small A-frame rooms with two twin beds in each, and a shared bathroom at the base of the treehouse. We all unpack, eat some dinner, and light a campfire before climbing up into the trees.

I'd tried my share of marijuana in high school and learned quickly it wasn't my thing. Even ended up quitting at just 16. Haven't smoked it in almost nine years. But tonight, after our adventures picking up Zulu, pushing the car, our clay-covered married couple, and of course, Big Mama's village, I *am* smoking this in the African Safari Tree House.

We pass around a couple of joints and laugh and laugh at the crazy fun of our adventurous day. It's so hot in the tree houses that we all drag our mattresses onto the connecting patio, an impromptu slumber party.

"I'm in AFRICA! We should go for a walk." It finally hits Tristan too.

"Yeah!" we all agree, his enthusiasm contagious. I grab my teeny-tiny pen-light as we head out. We're all walking, talking, smoking, and laughing. It must be about 2:00 a.m. now and we're having the time of our lives. We walk a mile and a half before we hear it.

SNAP!

It slowly dawns on each of us.

We've just heard a stick break. Something's nearby. Probably large, probably hungry. Probably like the rhino, buffalo, elephant, leopard, and lion. Oh yes, we're *inside* the Hluhluwe National Park.

A bunch of stoners who've forgotten this crucial fact.

Instinctively, we form a circle with all our backs together, while I slowly raise my pen light to the bushes. We're all sort of shifting around in a slow circle, like some crazed Riverdance, just our feet moving while we have our arms braced, ready for anything. The pen light reflects *hundreds of eyes* sparkling back at us in a 360-degree circle. Eyes of all shapes... and sizes... and... *heights.*

We collectively gasp. We're surrounded by God only knows what, and God only knows if they've eaten a meal today.

Again, the familiar call "*RUN!*" this time from my very wise friend James.

By nothing more than moonlight and a crappy pen light, we *run* back to our treehouse as fast as we can. We climb that ladder like our shoes are on fire. The night ends in shaky laughter, but laughter all the same.

I'm drifting off to sleep and I hear it. No doubt about it. A male lion's loud, guttural, very primal roar. I'm so very, very, *very* glad we're in the tree house, and this roar is not for us.

I notice the insane level of deafening silence for a good several moments following the roar. It seems everything from elephants to baboons, even the crickets, freeze for a moment to see if they have angered the most beautiful, most dangerous of them all.

With not another sound from the lion, slowly the bugs start with their chirping and the rest of the night joins in.

I'm on a real-life African Safari.

The morning begins much earlier than anticipated with a small downpour. All of us are sleeping on the mattresses we'd moved outdoors onto the patio. No one else is bothered by it, so I let them sleep in a bit more while I wrap myself in my blanket, covering my hair like a hijab, and go for a walk through the trees. The same trees that seemed scary last night seem peaceful and amazing in the glow of the morning. In my imagination, lions wouldn't attack me in broad daylight.

Jarring the morning silence, there's a small honk behind me as I turn and see the car full of the fellas. They're videoing me, laughing that I'm the first wild animal they've seen on safari. So, with some tea and donuts (Yay! No rice!) we begin our safari properly, in the safety of a car this time.

The animals, all friends in the wild, go to the watering hole at the same time, looking out and warning each other if there's danger. Families of hippos in the river, zebras, giraffes, baboons, kudu, and more elephants, of course. I discover the most dangerous animal is not the lion, it's the hippopotamus. Who knew? They're large, angry, territorial, and very clumsy.

We're feeling adventurous, so Tristan and I decide to sit with our bums on the edge of the windows, our legs inside with our heads and torsos hanging out of the car while we spot a herd of water buffalo. Two of them notice us and even begin to chase us while trying to keep pace with our car. They get awfully close. We both sit back down in a hurry so as not to spook them. *Though they're really spooking us.*

We see a lone wild rhino eating behind some bushes. I decide this will be a great photo op.

"Hey, James, can you turn off the engine and we'll see if he comes out more from behind that bush?" I ask in a whisper since our windows are down.

"Sure," he says.

We wait there silently for 15 minutes until I decide to slide out the window. From my seat, all I've seen is big rhino butt. If I could just get around that little corner, I could see his face.

"Just trying to get a quick picture!" I whisper to the fellas.

"Maybe not a great idea!" says Tristan.

"Yeah, get back inside, Pam!" whispers New Zealand.

I see James frantically shaking his head *No* from the driver's seat. But I can just see the rhino clearly from this angle, his hind quarters are 20 feet in front of me.

Ignoring everyone, I slide around the edge of the car to try to capture his face. I step slowly but, obviously, he can hear me. The rhino turns slightly and looks at me with a few branches hanging from his mouth.

Perfect shot! I hold up my camera and take the quick picture. As I hit the button, instead of the usual "click" sound, it seems this is the last picture available on my film-loaded camera. The camera decides to *rewind* the film with a loud, hunkering, clamoring noise. Sounds like I'm taking off in a jet during this peaceful moment of my rhino friend's lunch. He doesn't like it *or me.* He takes a large step toward me with a loud grunt and a threatening scrape of his front foot, like an angry bull to a matador in a Spanish bullfight.

I'm *petrified.* I'm the dummy who wouldn't listen to the others who unanimously thought my being outside the car with a dining wild rhino was a Bad Idea. All the fear from being chased by that elephant herd comes rushing back. *What the hell am I doing?*

I react. I take a huge step backward and dive sideways headfirst into the car's open window. The ass of my jeans rips wide open. *Yep, this is happening.*

James is way ahead of me and has started the car, flooring it in reverse away from the fat angry unicorn. My arms are splayed out on the car floor, with my legs and bare cheek hanging out in the safari sunshine. The fellas help pull me in as we speed away. The rhino charges us a few steps, loses interest, turns, and begins to feast again.

How unexpected that my camera would be extra loud in that one instant. So worth the shot, even though it's my only pair of jeans and they're repaired that night with the only thing I have handy. Duct Tape. Yep, The Wild.

"Hey, isn't today your American July the 4th?" James asks me.

"I guess it is!" I reply. The rest of the evening turns into me teaching them our American anthem and other hilarious traditions. Obviously, the Brady Bunch theme song comes up and teaching American government and culture with School House Rocks songs: *'I'm just a bill, sitting here on Capitol Hill...' 'He puts a feather in his cap and calls it macaroni...' 'Purple mountain's majesty...'* They love it all.

"Are your mountains really purple?" asks Tristan. Or maybe they're high again.

*****

After a wonderful road trip with my new buddies, I spend the last few days with Camryn, Vanessa, Evie, and Charli. It's time for me to say goodbye and head back to Los Angeles. I hitch a ride with some other travelers to Jo'burg for my flight out the next day.

I'm not ready to go home yet. I'd begun the trip with $3,300.00 but in three months abroad, I've only spent $900.00. Buying and selling a car was totally worth it. Wow, eating white rice every meal, morning, noon and night, *every damn day* really has paid off. Okay, okay, I snuck a snack a few times, but stuck with it 95% of the time. Meeting so many pals to share costs and especially the hospitality of the locals, I'm in LOVE with this world. I buy a small turtle statue souvenir.

I arrive at a hostel in the middle of the night, making a quick call home. "See you in less than 48 hours! Love you Mom!" I hang up with California. My flight out's early tomorrow morning. All my money is in South African Rand. The front clerk tells me the money changers will still be closed at the airport and it's illegal to take more than a few hundred Rand out of the country. Like *jail* and *smuggling* illegal.

I hadn't thought about that. If I just went for it, then, at best, the customs agents would confiscate the majority of my money. The very worst-case scenario, I'm in jail. Not sure what to do, the front desk guy's playing pool with some friends and we all start talking about my situation.

"Could I, like, buy some diamonds tonight?" I ask.

"Well, yes, but you can only legally travel with so much as well... same problem," he says.

"Could I mail it home to my mom?" I ask, trying to think.

"African mail? You'll never see it again, I guarantee that!" he says, twisting blue chalk on the cue stick.

"Can't help overhearing," says a male French-Canadian accent from the bar overlooking the pool table. "I just arrived today. What if I trade you my American Dollars and buy out your Rand?" Yes please!

In the dark night light of the already closed front office, the few of us count out the exact amount to trade and I'm back in action. There really are angels. Thank you, mysterious stranger.

*****

Just a year earlier I was living in San Diego, California, and my fully loaded moving truck had been stolen. I was late to work and parked the truck on a side street in La Jolla, the fancy part of town. With a clear repetitive thought *"Do Not park everything you own on the street!"* I argued with myself *"It'll be fine!"* and ran in to explain that I was late. I was handed a glass of wine and told I could take the night off. When I came out, my truck was gone.

Nothing was ever found, and I couldn't get out of bed for a week. I'd not only lost material things, but my memories, my identity. My "thought" had been my *instinct*. I was devastated that not only had I lost everything, but I hadn't recognized my own gut instinct. I wouldn't make that mistake again. Only now, I begin to see it might not have to remain a tragedy.

I don't have a husband, kids, a mortgage, or a job, even a car or anywhere to call home. No furnishings, clothing, photo albums. It was *easy* to buy a backpack, stuff a few new pieces inside, and head into the great outdoors. In my new understanding, this is the very definition of freedom. I'm going to make the best, coolest, most fun memories that I can enjoy in old age.

No one can ever steal my memories from me. I'm free.

*****

That next morning, I enjoy my last "bunny chow," which is a South African traditional meal of an entire unsliced loaf of bread with the center pulled out and filled with things like soup, stew, or beans and cheese. Probably for the best I didn't discover these early on.

I board my flight home to Los Angeles, with a stopover in Kuala Lumpur, Malaysia.

Walking onto the flight, I keep thinking about the things I've learned and the people I've met. My only two regrets are not seeing a wild lion and not being able to go to the Drakensberg *(Dragon's Tooth)* Mountain ranges with its *frozen waterfall* that you could hug. I'm exhausted, yet already nostalgic for this wonderful country.

I really just want to continue traveling.

Malaysia

Kuala Lumpur

- Black Hole Sun -
Soundgarden

"Hi, I'm Tony!" says my flight mate as I sit.

"Hi there, I'm Pam," I smile while we shake hands.

Happy to meet a new friend for the next 15 hours. We talk and talk and talk. He's never left his home of South Africa, so Asia will be his first time overseas. He's asking me for advice, like I'm somehow a "pro" at this now. I'm completely jealous that Tony's journey is just beginning.

Halfway there, after a long sleep where I dream of myself in a '*Romancing the Stone*' type adventure, I'm feeling bold, not ready for home, with plenty of money in my pocket.

What if I just keep going? Am I allowed to keep going? Well...I suppose it's now or never.

I show my ticket to the nearest stewardess.

"Excuse me, can I get off the plane in Kuala Lumpur?" I ask, nervous, yet hopeful. My fate is in her hands. If she says yes, I go alone to Malaysia. If she says no, I go home with $2,400.00 burning a hole in my pocket.

"Yes, sweetie, you bought a ticket that's good for a *year*!" she answers with a smile.

Oh, great joy.

After a lively yet uneventful flight, Tony wakes me and points out the staggering sunrise as we descend into Kuala Lumpur, Malaysia. From the airplane's tiny window, we see we're above a solid blanket of bleached cotton. A hint of the rising sphere of bright red sun, just peeking above the soft, rosy, layer of clouds.

We seamlessly pass through the thin layer of veils with the lower half of the neon crimson sun hanging down from under the top layer. There's a second, lower layer of milky, cotton-candy clouds, each pocket filled with blood-red pools of sunbeams. It's shooting real-life laser beams of scarlet into the pockets of this cloud sandwich above and below us.

Tony and I are both speechless at this spectacular vision. As we break through the lower layer, we see miles of misty jungle surrounding a city of huge skyscrapers.

My new adventure. And it looks '*Good Morning, Vietnam!*' hot.

We swap addresses as he goes to meet a connecting flight. I'd better call my mom because she's picking me up from the LAX airport in about 17 hours. I explain I'm staying for another few weeks and will keep her posted. Got to love Mom, she encourages me to have fun and be safe.

Outside, the humidity smacks me in the face.

Looking for a bus, I bump into two girls from New Zealand—Tina and Ashley. We all decide to catch a taxi into the city and share the fee. Once they hear I don't know where I'm staying or where to go, they invite me to come along. This traveling *solo* thing is really working out. I haven't been alone for five minutes since I started traveling "alone."

As we hit the main highway in the taxi, we cannot believe our eyes. Kuala Lumpur is crazy. A million cars, buses, motorcycles, mopeds, bicyclists, rick-shaws, skate boarders—*I swear I even see a hoverboard—no I don't*—and pedestrians all trying to get through the same traffic light. At the same time. On a red. With constant honking, all the cars driving so close you can even ask if the person in the car next to you has any Grey Poupon mustard *(but, of course)*.

None of us have been to Asia before and it's like a circus to our senses. I just don't know where to look next. Everyone has food stands, cooking things like snake, fish heads, even chocolate covered crickets. Or selling tapestries, baskets, music, and gems. Everyone smokes and wears flip-flops. I spot a dog on the back of a moped sitting up on the seat, the driver weaving through traffic, the little white fuzzy face focused on leaning to and fro. I point it out to the girls, and we share a giggle.

My two new pals are on their way home from Africa and headed home to New Zealand, with just a few days in Malaysia. They tell me about their planned excursions.

"Tomorrow we're going on a tour of a rubber tree farm and a bug farm that ends up at some cave temple. Want to come along?" Tina, the ginger, says.

"Count me in!" Who doesn't want to go to a rubber tree and bug farm? Hoping this is a *little* bug farm, not your *big* bug farms.

After freshening up, we enjoy a night on the town, grabbing some dinner of WHITE RICE, ugh! At least with vegetables and sauce this time, and drinks—watermelon juice served in a plastic baggie with a straw sticking out—while getting to know each other. Turns out they got chased by a herd of elephants too. We turn in early and all sleep for 15 hours.

The next morning, we hop on a tour bus and see the sights. First stop, the tree farm. For an hour we all learn to tap a rubber tree. Because who knows when I might need to make some sneakers, a basketball, or even condoms, right? A few practice-tries later and I've got real ooey, gooey, rubber dripping down my hands like maple syrup. Something to add to my resume, now I've acquired this random skill. So, I've got *that* going for me, which is nice.

Next stop, yep, your *Big* Bug Farm. When in Rome... I squeeze my eyes shut during half of it, and yet end up volunteering to hold a scorpion with oversized salad tongs. We pass several giant praying mantises, geckos, and snakes. We love learning about those huge, bright, turquoise butterflies. Malaysia doesn't believe in wasting anything if possible. It's even people's jobs to go out into the jungle to *find broken butterfly wings* to make into gorgeous collage art.

Next stop on the tour is the famous Batu Caves, just a short drive out of the city. Just have to climb up 272 dull gray steps into the Hindu shrine. In this humidity and heat, the climb is beyond excruciating. We three push ourselves, laughing about Girl Power, and how if we were being chased by elephants right now, we could do it. At the top, we find hilarious monkeys drinking Cokes and gorgeous statues of deities. The view of this sprawling city from the top of the steps is phenomenal. Jungle mixed with skyscrapers. The air feels thick, and the sweat drips from our faces.

It's a hint cooler as I enter the mouth of the cave. If only it could talk. There would be thousands of stories it could tell about those who visit here. I notice it's not dark in the back. Natural skylights direct more sunbeams onto the carved Hindu Gods, decorated with fresh flowers and holy water. I'm excited that this temple is still used today for real Hindu prayers, festivals, traditions. It's not solely a tourist attraction. We say some soft prayers while we lay flowers as it begins to lightly rain through the skylight. Rainbows appear momentarily, then *poof!* Gone. *A gift from God?*

After lots of laughter, I say goodbye to the girls the next morning as they continue home. I stay in our humble abode for another few days of exploring the city. While I'm changing money into Malaysian Ringgit at the bank, I notice a security guard seated in his chair. He's slipped off his uniform shirt so he's just wearing his stained white wife-beater tank top, his shoes slipped off and, best of all, he's *fast asleep*. I feel *so* safe if there's a sudden bank robbery. *Or maybe I should rob the bank?*

I organize a bus ticket south to Singapore for the next day. I think I'll head over to Indonesia as well. I'd learned quite a bit of the language already as I'd helped Jay practice his phrases, even some witty jokes.

I'm making my last day in Kuala Lumpur epic. I head over to the fanciest hotel I can find and act like a high-roller all day long. I guess royalty, prime ministers, and presidents stay here. So why not me? I hear about a roof-top swimming pool and head up in the elevator. Grabbing one of their pool towels, I set up camp for a few hours like I own the place. A gorgeous swim on a hot, humid, sticky day is just what I need.

"Something from the menu, ma'am?" A deep, accented male voice awakens me as the waiter hands me a menu.

"Sure, ummm, uhhhh," I stutter, as I already know I'm causing trouble and will likely cause more. "How about some orange juice?"

"Right away, ma'am," he replies. "And what is your room number?"

"Number 272," I say without missing a beat. "And a tuna sandwich?" I add. I feel like I'm in the 80's movie '*Fletch*,' where Chevy Chase is at the country club that he doesn't belong to and keeps ordering expensive things, telling them to "*Put it on the Underhill's bill!*"

"Of course, ma'am," he turns to walk away.

"Actually, can I get a slice of that chocolate brownie as well? Thank you," *I know, I know, I'm a cheeky bastard.*

"Right away ma'am," and off he goes.

After I've scarfed down a delicious lunch, I sign a scribble on the bill, dry off and *quickly* leave the building. Sorry, Underhills, you would understand if you'd eaten nothing but white rice for months. Here I am, making choices that might change the world forever and I wouldn't even know it. Maybe the Underhills are just rekindling their marriage after a huge affair, and when the wife goes to dispute the bill, she hears of a busty, bikini clad, young woman who billed it to her husband's room. That'll go over well. Remorse? Sure... *but comedy? Absolutely.*

And it wasn't totally lost on me that my fake room number had exactly matched the number of stairs that lead into the Batu Caves... 272. He's a local—he should've caught that.

"Do you know you have a black girl's booty?" I hear an American accent as I step from the lobby out into the humidity.

I turn, debating how to answer. *Yes, I know.* A huge black man in a tiny muscle tank top smiles at me, he's wearing a cowboy hat with long tourist beaded braids down his back.

"Oh, you're American?" I feign shock.

"You too? I'm from LA," he says arrogantly.

"Me too, just north," I smile.

"Oh yeah? Where?" he steps closer.

"By Magic Mountai—"

His hand flies up in my face as he interrupts me. "That's not proper LA."

You just made me so very happy I'm not landing back home yet.

"I'm traveling with my band. What are you doing here?" he winks at me.

"Just passing through, arrived from Africa, heading to Indonesia... just backpacking." I shift on my feet. I don't feel like meeting someone from LA.

He steps closer to me, throws his beaded hair over his shoulder, looks me up and down, licks his lips, "Look, baby, are we going to do this or not?"

I laugh. "We are definitely *not.*" I walk away.

"C'mon, baby, let's get you a nice drink!" he yells after me.

Ugh! Arrogant LA men. I don't look back.

I hop onto the bus to Singapore.

Singapore

Singapore

- Misunderstanding -
Genesis

I purchase an afternoon ticket and it's about 10 hours on the local bus until I'll arrive. My only style, so far, has been to arrive in the airport, go to a money changer immediately, and *then* figure out where to stay and what things to see. I seem to be the only tourist on this ride. Without Camryn or my New Zealand girls, this is my first real country where no one has my back if I mess up… *yikes.*

The bus lurches to a stop at 1:00 a.m. and I wake up, see everyone grabbing their luggage, and realize we must've arrived. I look out the window into total darkness. It's pouring rain. Cats and dogs, even horses and goats. I've been carrying a black trash bag to cover my backpack for just such an emergency. As I search through my pack for it, I see we're not inside a brightly lit bus terminal as I'd expected. We're not near a bus terminal at all. As a matter of fact, we're not near any city, buildings, farmhouses or even a streetlight. It seems *this* is the stop, on the edge of a muddy dirt road at a crossroads on the edge of a jungle. *Oh my.*

Well, I guess I'll just follow the other people. They'll be walking to the city as well. The driver has already left the bus, the 15 other riders are dispersing rapidly, and *no one is going in the same direction.* Each person is running into the darkness to get out of the rain.

So, here I am, standing in hiking boots ankle deep in mud. In a monsoon. My backpack protected by a thin, black trash bag, not even a hat for my head, and I must decide which of the four directions to walk. Into the pitch black, I see nothing, just a shadow of thick jungle. I've seen pictures of Singapore with its ultra-fancy hotel, Raffles, on the edge of a modern harbor and *this is not that.*

Eeny, meeny, miny, moe. I choose one "street" and off I go. A half hour in, I'm really questioning my choice. *Are there huge snakes in this jungle? How big is Singapore? Can I walk for days and not see anyone?* I'm so hungry and cold. At least I can squat down to pee on the road and no one'll notice. After walking about 45 minutes, I begin to see some lights, and as luck would have it, they are streetlights. Lo and behold, a row of fancy hotels. I literally hear the angels singing. I go inside and see a real person. He's a chubby Chinese man smoking a cigarette at the side door.

"Hi there," I say with a bright smile. "Do you speak English?"

"Yes," he says, no smile, with a thick Cantonese accent, "What you want?"

"Yes, thank you," I reply. "I've just arrived into Singapore and don't have anywhere to stay. Do you have rooms available?"

"$250 US a night," he states roughly. I guess he's judging me by my soaked appearance, mud splashed from the knees down, and my awesome trash bag.

I visibly flinch at the price. I have only US traveler's checks again, US dollars, and some Malaysian Ringgit in my pocket. "I don't have any Singapore Dollars. Are you able to change money for me?"

"Only for guest," he says gruffly and *walks away.*

Thinking quickly, I follow him, smile, and with my voice dripping in honey, say "Well, how can I check in if I don't have Singapore Dollars?"

He pauses while he thinks about the common sense of it. "Okay," he barks at me to follow him to the front desk. I slide in my muddy footsteps through the lobby. He changes $200 plus the rest of the Malaysian money and as he turns to choose a room key for me while picking his nose, I shout "Thank you!" as I run out the door.

$250 a night, yeah right, not with that kind of customer service attitude.

Now that I have some money to spend, I go into the next fancy hotel and there are two Chinese men at the front desk. The moment they see me, they rush right over, offer to take my backpack, and get me a fresh towel and a cup of hot tea. Now that's what $250 a night should feel like.

I explain my situation, that I'm looking for more of a backpacker's lodging. At once, they get me a phone and point me in the right direction in the phone book.

"Hello," uh oh, another rough attitude on the phone, I can tell immediately.

"Yes, hello," I say. "Do you have any rooms available?"

"Yes," he says. He then *hangs up the phone.*

I call right back.

"Hello," he says again, annoyed to be bothered to do his job.

"Yes, hello," I say with equal annoyance in my voice. "How much are your rooms?"

"Seven dollars!" he hollers and hangs up *again.*

I call back once more, and he shouts "HELLO!" I shout, "WHERE ARE YOU LOCATED?!" and as he's barking out the address, before he gets the chance to hang up, I shout "I'LL BE RIGHT THERE!" Yep, that's what seven dollars a night customer service will get you.

It's pretty far so the nice gentlemen call a taxi and off I go into the night again. As I arrive at the new hotel, the phone guy turns out to be the biggest, fattest Jabba the Hut in a torn, soy-sauce-stained, grungy white wife-beater shirt and his boxer shorts... picking at his bare toes. *Lovely.*

He grunts, grabs me a key, and I get to my "room" which is actually not as bad as I pictured. It's about the size of a small closet. As I open the door, it hits a lumpy twin sized bed. But there's a ceiling fan, and a locking door. *Ahhhh, heaven.*

I notice a clean but stained, so-thin-you-can-see-through-it, scratchy sandpaper hand towel folded nicely on the nightstand. Oh, yes, this is going to be my new shower towel. I get rid of my huge, thick, fluffy, beautiful, shower towel as a gift to this asshole. I fold up the new one so small that I might as well be packing gossamer. Now I literally have half of my backpack free again, for souvenirs and God knows what else this unexpected continuing adventure might bring my way.

After a shower at the shared "BATHROOM DOWN THE HALL!" I dream of people yelling all night.

My first morning in Singapore I'm awakened at 5:00 a.m. by the Islamic Call to Prayer voiced over the loudspeaker of the mosque that must be nearby. Like, in-my-eardrum-near. I just got to sleep at 3:00. After I wake up properly, around noon, I decide I'm going to visit the lovely mosque.

I'd always wanted to go to countries like Morocco and Egypt, so this can be my Africa away from Africa. I dress in my nicest clothes, a long black sarong, a loose black T-shirt, close-toed shoes. Okay, okay, my muddy hiking boots wiped off the best I could. *Hey, it's either that or flip-flops.* Plus, another sarong (to cover my hair and my shoulders) that I'll just carry with me until I arrive on site. There. Respectful of another culture, one that I am definitely interested in.

I walk down the street to the mosque and appreciate the beauty of the architecture. The rounded rooftop, the marble pillars, the carvings, and the Islamic writings. As I enter the gates, two men run over to me and start yelling and waving their hands. *Oh, GOD! What now?* I know, I know I must take my shoes off? Nope... you can see some of my hair? Nope... What? An offensive sexy ankle? Nope again.

They don't allow women to enter. *Ever.* I'm grumpy now, I have to leave.

"Sorry about my vagina," I mumble under my breath, ripping the veil from my hair.

I am *not* feeling Singapore.

Back at my room, I grab my pack of chewing gum and let loose. Growing up, all I knew about Singapore was that chewing gum was illegal. I go down to the street in my regular clothes and chomp, chomp, chomp that gum. Take *that* Singapore. I'm no stranger to danger. I might even blow a *bubble.* Just a small bubble. I don't want any lashings over Hubba Bubba.

I don't know where I'm going or what's cool in this country, so I end up just following locals down into what seems to be an underground food court in a crowded mall. I'm starving. And, yay, there are pictures of food for people just like me. I choose some noodle-veggie combo thing which tastes wonderful, but comes with a side of—you guessed it—white rice.

There must be 300 people dining in here and I'm the only westerner. I feel a little like I'm in the spotlight. People stare at me as I eat, but this is a big metropolitan city. Maybe I live here, no one knows.

I purchase a soda can with funny writing on it, so I don't know what flavor it'll be. I take a sip, swirl it around on my tongue and, ironically, it tastes like bubble gum. Yum. Not bad... I keep drinking. As I tilt the can back a bit, I feel chunks, yes, multiple chunks fly into my mouth. *Oh no*. Cockroaches in the can? Some other bug I can't even imagine? *Vomit?* I spit out the chunks and realize with a giggle that they're very tiny, and very stale, gummy bears. Still, I cannot continue with the drink. I choke down the rice with my meal and just as I'm finishing up, two Asian men come up to my table and say hello.

"Hi there," I say.

"Where are you from?" the tall one asks in perfect English.

"I'm American," I say.

"Me too! I'm from New York!" he says excitedly. Tim introduces himself. He's fit and about 30 years old, with his hair gelled back like a teenaged biker from the 1950's. His T-shirt sleeves are rolled up stylish. He introduces his boyfriend Tom, from San Francisco. Tom's also fit and stylie, with a manly square jawline. They're both Mandarin and have lived here for five years. We share a few laughs about travels. Okay, specifically, about my introduction to Singapore thus far.

"Will you come out with us for dinner tonight?" asks Tom. "We'd love to show you our cool spots and take you out for a great time."

"Thank you! Yes, please, it sounds fun!" I'm excited to meet new friends, plus gay guys won't hit on me.

That night, they arrive in a fancy car and off we go. We park and walk down to the Esplanade. It's *beautiful*. I feel like I'm walking along the Seine in Paris.

Lining the river's edge are outdoor cafés with brightly colored umbrellas over each table. Fragrant flowers and trees everywhere, inviting wild birds. Small boats float down the canal with smiling guests on board. Live musicians busking gently with a saxophone or an acoustic guitar, even a harp, far enough from each other to avoid overlapping. Each restaurant we pass is filled with finely dressed people laughing and pouring wine. We stop at one of their favorites and soon *we* are the finely dressed patrons laughing and pouring wine.

Now I'm feeling Singapore, for sure.

Funny how just last night I was tromping through the mud, in the rain, getting lost and getting yelled at.

Interesting conversations flow during a fantastic authentic Italian dinner. After gelatos, Tim and Tom suggest going to see the Mer-Lion.

"What's a Mer-Lion?" I ask, wondering if I heard them correctly.

"You've never heard of it?" Tim's shocked. I guess I'm the only traveler to just get to a place and *then* find out about where I am. "It's Singapore's most famous landmark, you've just got to see it!" He says, his eyes filled with laughter. Laughter *at* me I'm sure, but I'm going to pretend it is laughter *with* me. They pay for my dinner and we're off walking again.

There it is, a large statue just as he described it, the head of a lion and tail of a mermaid, with water spouting out of his mouth into the river. Glamorous. We pass an artist with his canvas on a tripod, a nearly completed Picasso style of the Mer-Lion.

Tim shares the history of how the country is an island and its original name was Indonesian for "The Sea." War in the fourth century destroyed Singapore but, later the land was rediscovered. In the eleventh century, the founders came across animals they'd never seen before, including a lion. (*Really? A lion this far from Africa*? I guess, since there are Sumatran tigers) Hence, the Lion and the Sea. Today, it has some of the most well-known harbors used for trade around the world.

As we're walking back, we pass a lantern-lit nighttime flea market and Tom wants to have a look. "We ***need*** refrigerator magnets!" I hold back a laugh at the urgency and the emphasis on the word "need." So, we're looking at fridge magnets and Tim says "Ooooooh, look at these ones, kind of retro 1950's advertisements." Tom says, "No, these ones with Chinese dragons on them match our kitchen for sure!" and back and forth it goes until it becomes a serious argument. *Awkward.*

They're bickering like an old, married couple while they keep looking at new magnets. For ages. Tom even starts sticking out his lower lip in a pout that would rival any true-to-life three-year-old temper tantrum. Then he grabs Tim's rolled up T-shirt sleeves and rolls them down. Tim's shoving him away and each time he rolls a sleeve back up, Tom unrolls the opposite one. I'm trying not to laugh. Finally, Tom squeals as Tim gives in and they agree on the Chinese dragons. I insist on purchasing them, so they'll always remember me and to say thank you.

The next morning, as I sit in the crowded harbor waiting for my ferry, a Chinese salesman spots me and in perfect American-radio-DJ-voice launches into his sales pitch. I can't get a word in edgewise. Finally, just to get rid of him, I say loud and clear, "I don't speak English." I smile. He stops, gives me a weird look, apologizes, and walks off. *Ahhhh, this could be useful.*

Seriously, loving Singapore.

Indonesia

Riau Archipelago

Batam Island

- Up Around the Bend -
Creedence Clearwater Revival

After only a few days in Singapore, with some hilariously bad and some hilariously good memories, I'm now buying a ferry ticket to Indonesia. Jay had been years before and had arrived here again after Africa, although he's back home by now. He'd given me some awesome information.

"They don't use toilet paper. They eat with their right hand and wipe their ass with their left." *What?*

"You might get a really hard **crunch** while you're eating and, believe me, you don't want to know what it is, so just think it might be an uncooked piece of rice and keep on chewing." *Eeeeeww.*

"Don't order any meal with mushrooms unless you want a Magic Mushroom day." *Good to know.*

"Don't get your hair braided, you'll look like a dorky tourist." *No problem.*

But also "Disneyland isn't the happiest place on Earth, it's Indonesia. Everyone's smiling all the time." *Okay, this one I can get into.* We'd practiced quite a few key Indonesian phrases, numbers, and food, I'd absorbed so much that I'm unexpectedly ready to rock.

As I'm in the terminal to board the small ferry that will whisk me away from Singapore to Batam Island, Indonesia, I meet two new friends from San Francisco. Kiersten's got a brunette pixie cut and hazel eyes. Her boyfriend, Chase, is sandy-haired, blue-eyed. He's in *Love.* He barely looks at me. I'm a bit envious of them traveling the world in love.

We drop our backpacks onto the X-ray machine for the agent sitting behind the monitor. When he gives an odd look, we realize no one else is doing it. Yes, it's his job, but he's not enforcing it as all the locals just walk past. The monitor isn't even plugged in. We bond with a laugh at ourselves and our rigid rules instilled in us from home. We board the boat and get comfortable for the 45-minute ride.

"We're only in Batam for one night, then we catch a flight to Borneo." Kiersten's excited to be going back. She'd been a few years earlier to see the orangutans at the protected farms. She's a waitress and it took her a few years to save up, but she knew in her bones she'd be back as soon as she could. This is Chase's first trip overseas, and he's equally thrilled to get there.

Somehow, we get royally ripped off splitting a taxi. Plus, he drops us off at the only fancy hotel in town. We agree to split one room, $20 for the night. As we're all freshening up, Kiersten takes out a few huge bags filled with batteries. Big old batteries, small batteries, literally a Sav-On drug store full.

"What's with all those?" I ask.

"Last time I was in Borneo, I took so many pictures, my camera ran out of batteries. There's not exactly a store nearby to pick up more, right?" she says with a laugh.

I grab one of the bags to test its weight. Just this one bag must weigh 10 pounds. She has *four* bags of batteries. Plus, her backpack is HUGE. It goes way down by her knees and way up above her head.

"Oh my God!" I exclaim. "How much does your bag weigh?" She admits almost 95 pounds. I'm so grateful mine is tiny. It weighs 25 pounds—even less without that mammoth beach towel.

Chase pipes up, "I'll carry the batteries for you in my pack."

"Are you sure?" she can't believe it.

"Of course!" he says with a smile, as I'm envious again.

We throw on our bathing suits, ready to find a bite and a drink, then hit the local beach. Chase heads to the bar to grab us a few beers and Kiersten tells me they're newly dating and while she likes him, she's glad he's here, she isn't feeling things as quickly as he is. He'll win her in the jungle, I'm sure of it.

After lunch, we head over to a small beach cove and jump into what we think is the ocean, but we're walking into the slimiest mud pit. It looks like lake water with tiny wave ripples, but nope, somehow, it's all mud. Yet, no one wants to stop this. We're all miserably hot. It's sticky-humid. We wade in deep, up to our waists, in this cool, refreshing muck. We sit and chit-chat in the filth for about an hour. Hey, people pay big bucks for the mud wraps at a day spa, right? *Yet... maybe this is how you get worms and parasites in tropical places? Oh, God, please, no.* At this realization, we agree to shower up to go out for dinner. She even has nail polish and we do our toes.

They tell me all about Borneo, how it's a real river highway place, and they're heading in during the beginning of the rainy season. They're only there for two weeks, would I like to join them? I think about it and decide no, let them bond. Plus, if the rain hits hard in the beginning, that means the river floods and you could get *stuck for months*. Fun with a lover, not fun alone.

I tell them all about Africa and my Asian adventures so far.

"Hey, beautiful! I just heard there was a white girl on the island!" We hear an Australian accent.

Kiersten and Chase are all cuddled up, obviously a couple, and I realize this must be a call to me. I turn to see three western men walking up to our table. We introduce ourselves, they're all living and working on the island. Jimmy is the loudest, funniest, not-shy-at-all-cat-caller. Tall with sandy hair and blue eyes and a contagious laugh. His friends are Greg, my height with dark brown hair, and Bungee Jump Benjamin, a stocky blond that runs the—you guessed it—bungee jump crane for tourists at Waterfront City Resort. We all chat, and once Jimmy finds out my friends are leaving, he offers a couch to sleep on at the home he shares with Greg. Yes, please!

Next day, I say my goodbyes to my pals and hop into Jimmy's "ute" he calls it, his utility truck. More like a small flatbed truck in the US, the kind you rent by the hour at Home Depot. He drives across the island, pointing out amusing street signs. One picture-worthy sign has nothing but an

Exclamation Point "!" With no curve, dip, cliff, speedbump, or anything to truly warn you about.

He pulls up to a modern condominium complex. Bright and sunny, air-conditioned with a private balcony, his place is tiled and fully furnished, with a full kitchen and even a pool. Not bad. He and Greg want to take me around the island and teach me about Indonesia, and I realize I couldn't have stumbled onto a better group of new friends.

Jimmy asks his cleaning lady to do all my laundry. She hand-washes everything and even waits on the balcony for everything to dry so nothing gets stolen. Then she irons everything, including my panties. It feels awkward, being so well taken care of. But she and Jimmy insist it's no problem. They tell me about tropical mites, the small bugs that can live in your clothing in this kind of humidity, killed easily with an iron. *Eeewww!* I guess my wash is *well* overdue. My new home-sweet-home for a full week and I can't believe my luck.

Greg, Jimmy, and another blond, tall Greg—Big Greg—are working abroad for a couple of years as electrical engineers on Batam. Their company bid for the job and once the Australians arrived, they let go of all the Batam islanders, bringing in other Indonesians to complete their team. Angered by this, the fired group of a hundred men came back the next day, armed with machetes, and *slaughtered all the men* who had stolen their jobs.

"Are you serious?" I'm completely shocked. "I don't remember that on the World News!"

"Well, it didn't even make Indonesian news," Jimmy laughs.

Little Greg nods his head. He explains that Indonesians aren't allowed to hear their own country's news. He proves it later with his radio. "*Here's the news in Thailand: blah, blah, blah...*" "*Here's the news in Singapore: yada, yada...*" "*Here's the news in Vietnam: whoop, whoop...*" "*Here's the news in Indonesia:* ________" Dead air. Radio silence. For three full minutes. Then on to the next country as if nothing had happened at all.

"So, weren't you scared the machete guys were going to kill you too?" I ask, still in disbelief.

"Nah, they just wanted their jobs back," Jimmy says, sipping his Bintang beer.

"What was their first day back like?" I wonder.

"Their first day they had to clean up all the body parts. Then we all pretended nothing happened and got right to work. Indonesia is fuckin' weird! It's absolutely fuckin' hilarious!" Yeah, *hilarious*. I'll remember to stay on their good side.

Jimmy points to the guy that valet parked our car outside the bar just moments ago. "See that valet guy? He doesn't work here, or for the city. He just bought himself an orange striped construction jacket and the other day, I pulled up, he took my keys and charged me ten cents to park. Been here every day since!"

Over the next few days, I feel like a queen. I'm driven around the lush jungle island to see the sights. Jimmy takes me to a small village, hinting at a ride on a yacht. He'd just told me of his ex-girlfriend who had been living here on the island. They'd ended things amicably when she earned her captain's license with a new job on a wealthy family's yacht. I assume she's in town and we'll all spend the day together.

"She's called *'Buy-A-Newie'*!" We're standing in front of a rickety little rowboat. My assumption couldn't have been more wrong. Seriously, this was built in, like, 1935 and has been patched beyond recognition. He continues, "...because I shoulda *Bought A New One*!" Giggles all around as I'm loving these Australian abbreviations for everything. We climb aboard and he rows around the edge of town, which is made up of 25 stick homes, each about 100 square feet, way up on stilts. No stairs, only tiny stick ladders to climb up, nothing but sand underneath.

"You could buy a home here outright for about $300 US!" Jimmy tells me. "See that one painted just there? That's fancy, maybe $400 for that one!"

There are children everywhere, some naked, all with huge smiles, playing with nothing but dirt. They wave and shout, "Hello, Mister!"

"That's the only English they know!" Jimmy laughs. He hands the kids some chewing gum and they all squeal with delight, immediately sharing. *And litter the gum wrappers onto the ground, which is already strewn with trash.* Jimmy explains that trash is a source of *pride* in Indonesia. If your family is wealthy enough to afford M&M's, you'd want to leave the wrapper in front of your house, so all the neighbors know you're rich. It's always been their way, but now the litter is all plastic, unfortunately.

One kid runs away and climbs up the ladder into his home and within 10 seconds, we see something that looks like rocks falling from the small hole in the floor of the home. "Yep," Jimmy explains, "that kid went in to take a crap." That is *crap* falling from his *ass* through his *kitchen floor* and onto the *play area* where all the kids are *barefoot.* As my face distorts into a "*GROSS!*" look, he explains high tide will come in soon, flood the area, and wash everything out to sea. *Lovely. Then it's okay.* At least I know what this island smells like now.

After the boat ride, he takes me to meet Vic, a South African who also owns a boat here. The big talk on the island is Vic's dock got new lights. Everywhere we've been today, Indonesians are shouting in very broken English to Jimmy "You hear Vic's lights?" "Vic new lights?" I'm guessing this is a small town.

"Wow, this is a small town, hmmm?" I ask.

"Yeah, that's how we found out about you and your friends," he laughs. "The ferry ticket agent saw you get off the boat and told the dock boy to make sure I knew about you. We *never* see white girls on this island. It was quite big news. We quit work for the day to look around for you. Every Indonesian who saw us coming didn't even ask what we wanted, they just pointed us in the right direction until we saw you!" We laugh and laugh. He decides to take the whole week off.

Jimmy takes me jet-skiing with the Gregs, with Big Greg's wife and kids. What a beautiful family. They've just arrived from Australia last night for a short visit and are thinking about living here for a year. I don't know if I could live here full time. But I'm in adventure mode, not marriage mode, so I'm glad I don't need to make the choice myself.

I'm roaring with laughter on the jet-skis. We're spraying the other friends, even some local bystanders on the dock. "Too bad you ain't having any fun!" Jimmy says sarcastically. We park the jet-skis and have a few beers with lunch at a bar near the end of a rickety old dock. Off in the distance, the buildings of Singapore are lit up like a mirror by the setting sun.

A rusty wobbly ladder curves over the edge of the pier, inviting me to climb down into the ocean for a quick dip. Jimmy peeks over the edge and asks, nonchalantly, "Hey, Pammy, you know the most poisonous water snakes in the world live here, right?"

"Ummm, no I don't!" I quickly climb back up.

He just laughs and insists on buying me a Singapore Sling drink when he hears I didn't get one while I was there. I don't think he's stopped laughing since we met. Maybe he's drunk as a skunk, or just hilariously Australian.

As we're leaving the bar, 40 Muslim teenaged schoolgirls with their hair covered want to practice their English with me. Oh, the fun of immersing myself and giggling with girls from around the world. They all want to know if I'm married and have children because I'm so old already. At 24. That's the beauty of struggling with a language barrier. You have to just say what's on your mind, not "be polite" about it. I love the realness and feel like I've forged new friends. I just hope they know I find them as intensely interesting as they find me, the stranger in their strange land.

Later that night at dinner, Jimmy introduces me to some Singapore ladies, who come over to hang out with them occasionally. "They're hookers!" he whispers in my ear.

Hmmm. Maybe I'd have been a hooker if I had been raised on this side of the world? No one ever knows until they live in another's shoes. He explains for years it sucked to be born a girl in China. Millions were drowned due to the 'One Child Per Family' moratorium on the overpopulation problem of one billion. Some other Asian countries rejoice at having a baby girl so they can eventually sell her as a prostitute.

How absolutely tragic either way.

I feel blessed to be from my home in California where I was able to play with Barbie dolls, play baseball, ride my bike to the park, and go to school.

We head to a bar that has international flags up on the walls from 70 major countries, but I don't see the USA flag. When I ask about it, the bartender tells me they took it down four years ago when some USA military guys were over-the-top obnoxious. *Oh no, I hate stories like this.* For a country as large as ours, not many Americans travel overseas, unlike the British. If a few military guys act up, those might be the only Americans they've ever met.

Before we leave, I've ended up making such friends with everyone that the barkeep goes into the back and gets the American flag and puts it back up, just for me. I've done my good deed for the day. The guys are proud of me and try to sing my national anthem, failing miserably, while the hookers are cheering for me. I feel patriotic so far from home.

I've chit-chatted with the girls all night and they're just lovely people. I notice none of the guys hook up with them before we drop them at their ferry. It seems like genuine friendships. On the way home, Jimmy thanks me for being so nice to them. He thinks Australian girls might have ignored them.

Jimmy's also house-sitting for friends and must check on their place while they're out of town. We head over to another fancy condo complex and inside their place looks like a Moroccan castle. Terracotta tiled floors with cream shag carpets, huge purple and maroon throw pillows everywhere to lounge on the ground with gem-colored draped curtains, low tables with orange candles, intricately carved statues. And travel books. I see a wall cabinet filled from floor to ceiling with Frommer's Travel Guidebooks to what seems like every single country in the world. Jimmy mentions that his friends are the authors of these fantastic travel book series, competing only with the Lonely Planet's Guidebooks. This revs me up to keep on traveling.

Today's my last day on Batam. Jimmy has babysat me long enough and he's only hit on me about 50 times.

"Not even a pash?" he asks, as though it's a shock to him.

"Not sure what a 'pash' is?" I wonder aloud.

"A snog?"

I shrug, I still have no idea. He does his best Texas drawl, "A make-out session?" He laughs at getting rejected again. "Anyway, I'll be in Jakarta in a few weeks if you want to meet up there?"

"Do we have to 'pash' if I agree to meet up?" I say with a wink.

"'Course not! But I'll ask ya again anyway!" he laughs. He's treated me like a little sister out in the world for the first time. I've learned some key phrases, "*Tidak Apa-Apa*" means "*No problem,*" "*Hati-Hati*" means "*Caution,*" and the most valuable one "*Ma'af, jenang gangu saya, saya ada pacar!*" which means "*Sorry, please don't disturb me, I have a boyfriend!*"

Bintan Island

- Hotel California -
Eagles

After hugs goodbye and a promise of where and when to meet Jimmy in Jakarta, I have my ferry ticket to the neighboring island, Bintan.

I plan to head south to Singkep Island, then it sounds easy enough to get over to Sumatra and keep going south through Java, then Bali, then if there's enough time, on to Lombok—maybe Komodo to see those Komodo dragons in real life—who knows? What freedom to choose. And so far, Asia's dirt cheap. I'm living on about $6 a day. Indonesia has a strict two-month visa law, if I break it, I *go to jail.* By then I'm sure I'll be homesick and return to Kuala Lumpur for my remaining ticket home.

There's a long line of locals getting onto the boat, and at 5'7" I'm a head taller than everyone in the entire country. Everyone's pushing and shoving, but I hang back and politely wave my hand to show '*After you, please, I insist.*' Everyone mosh-pits forward until I'm the last person on the boat. Half a seat's left covered in vomit. There's my answer, if pushing and shoving is the norm, I'll never be stuck with the vomit seat again... that I can guarantee. *When in Rome, indeed.*

After standing for the full trip, upon arrival, I spot men on motorcycles near the terminal, yelling out to see if I need a ride to a hotel. Jimmy said this is the taxi system on these smaller islands, and they're legit. I hop on the back of a motorcycle—no helmets— and grab the Indonesian stranger around the waist. *I'm hoping you're not a rapist, or in a gang, or a kidnapper who forces me to become an international drug mule... guess we'll see!*

We go halfway around the charming island, passing palm trees, thatch huts, markets. Families work in the yard with happy helping children, chasing after our bike, screaming "Hello Mister!" These little peanuts warm my weary bones.

We pull into what I suppose is a hotel. It's more of a hut/bungalow grouping around a small courtyard, like a tribal copy of the setup in Jeffrey's Bay, South Africa. I pay about $2.50 for the night. I'm led to my room with a mattress on the floor, mosquito net, a fan and candle on the side bench, with a broken wooden door and a tin roof. A shared bathroom's out on the edge of the courtyard.

I lay my bag down to go check it out, and lo-and-behold, my first sighting of the squatty potty. Yes, a cement room, with a slight curb running down the center of the floor. The 'toilet' on one side, 'shower' on the other. I use both terms lightly. The toilet is a cement hole in the ground. The end. That's the end of the description. There seem to be very faded, slightly grooved areas on either side of the hole for your feet. *How many people accidently step in the toilet each day in this entire country?*

The shower's a drain in the floor next to a huge cement barrel filled with water nearly to the top. There's a small shelf with what looks like a large soup ladle, like a scooper. I get it now. The barrel is the fresh water that no one touches, you just scoop out your share and pour the scoops over you for your shower. You use the scooper to fill your left hand to wash after using the facilities. *Um, I'm so glad I've got a packet of Wet Wipes.*

I'm traveling solo through Indonesia. I feel like Indiana Jones. Zero English is spoken here. But I get a free banana pancake breakfast in the morning and sweet tea all day long.

I have one of the men take me on his moped to the open marketplace, where I find fish, fruits, veggies, and hand-made jungle trinkets. Entire families work their own booths, with grandparents sleeping on floor mats behind each table, little kids running through the bazaar. I see massive mangoes, five times larger than at home. I never had exotic fruits before. There are fresher-than-fresh pineapples, mangoes, coconuts, starfruits, dragon fruits, and lychees. I grab a greenish mango larger than a football, bottles of water, a few crackers, and head back.

That evening, my hosts are preparing food for a potluck barbecue at the neighbor's house if I'd like to join them. Yes, please! I bring my mango to the potluck and help the women in the kitchen as best I can. One of the men at the party speaks a hint of broken English, and when he finds out I'm from California, he whips out a guitar, plays the Eagles '*Hotel California*' just like our barbecues at home. As they sing the words they don't fully understand, it reminds me I know quite a few songs in other languages that I equally enjoy... *La Bamba, Frère Jacques, La Vie en Rose...*

They hand me a warm banana leaf to unwrap. Inside is a starchy paste square thing, like a coconut gel baked tofu. It's okay... chewy, slightly sweet. Next, I receive a shellfish thing to eat, and hey, I like oysters on the half shell, so I try it. Big mistake, big, *huge.* It's beet red and overly juicy, with the texture like a tongue, and it tastes like—*gulp*—if I get a bloody nose that runs down the back of my throat. *So. Gross.* I can't go wrong with the fruit I brought.

A girl hands me a knife to slice my massive mango. It could feed 23 people, by the way. I smile and slice it right down the center, like it's a cantaloupe.

The women giggle at me as I start giggling too when I hit the large, unmistakable pit in the middle. They take it from me and show me how to slice down the sides a bit more. Hey, how am I supposed to know? Everyone politely passes on the fruit I've cut up, but I decide to try it. *It's gross as well.* Like an ultra-green banana, it's chalky and sour. Ohhhhh, it wasn't even close to ripe. I've got so much to learn, I'm like a baby.

The Indonesians are warm hosts and try to make me feel included. Everyone wants to know where my husband is. I guess it's pretty rare to have a girl going around the world by herself. In broken English, the gentleman translates that he and his friends think I'm beautiful. I literally look behind me as I think he's speaking to one of the other local ladies. I'm genuinely confused. I haven't washed my clothes in a week, and I have zero makeup on, my mop of curly hair rolled up into a bun to avoid the frizz of the humidity. I smile warmly and say "*Teri makasi,* thank you."

I get home to my mosquito net, curl up under the thin sheet, contemplating how I've never felt so beautiful in a lifetime in Los Angeles. I was surrounded by competing girls in high school. It became a fashion show and I wouldn't be caught dead without makeup on. Sure, people in LA always compare you to which movie star you might look like, and I'm proud to say I always heard Cindy Crawford the most, with Raquel Welch a close second. But this time it was different. I was not trying. I was just being me, the "*me*" God made, for real, no bells, no whistles.

I think back to the rotten dates I'd been on where men actually told me I'd be even prettier if I'd just fix the bump on my nose, remove the mole on my face, and I'd even been offered a free boob-job. *I'm already a natural DD. So...no.*

It's this kind of thing that can make a girl feel lousy about herself. This is the first time as an adult woman *I just fully accept myself.* My unruly hair, my beauty mark, my curvy figure, my "*ME.*" In this moment, I've reached another kind of freedom, one free of insecurities and free of jealousy, free of self-doubt and free of self-criticism. Free of the mind-reel that only I've been consistently telling myself all these years.

I believe him that they thought I was beautiful.

And just like that, I believe I'm beautiful.

In the morning I meet three other backpackers, two girls and a guy. We introduce ourselves and they're American as well. Savannah, Joy, and Jared are all from San Francisco, traveling around for two weeks before they begin their two years teaching English in Sumatra. Love these guys already, and mentally noting I need to check out San Francisco when I get home as these have all been my kind of people.

Savannah's a real beauty with curly brown hair, perfect olive skin, making her hazel eyes glow. I notice right away she doesn't shave her legs or underarms. She's even cooler for not caring. Joy's cute with black hair with cut bangs, bright blue eyes, and a sprinkle of freckles over her nose. Jared's a tall, lanky guy with dark hair and glasses. Maybe a slight Star Wars nerd, he seems like he knows what a lucky guy he is to be traveling with two adventurous ladies.

We share our life stories. They tell me how they all signed up separately to teach English and met just a week back along the training route. We all discuss the differences in our styles.

"Someone organized our flights and our host families picked us up at the airport when we arrived last week," Joy explains.

"The host families teach us a bit about Indonesia's culture, specifically Sumatra," says Savannah.

We agree it must've been hard to make a two-year commitment to a place you've never seen, but how easy it is to have those things set up for you.

"I've just started traveling a few months back," I share. "And every single day I have to make tough choices that sound easy. Do I eat that weird food, or what if I get deathly ill, all alone somewhere? Do I take a ride with that strange man? I hope it goes well! Even leaving my backpack while I shower, will it be stolen while I'm gone? But if I don't like a town I can move on tomorrow. Even that sounds like an easy choice, but what if I miss out on something cool by leaving too early? I don't have a credit card. I'm spending my life savings day by day here. I hope it doesn't run out. Hope I'm better at math now than when I was in high school!" I pretend to chew my nails like I'm nervous.

It's easy to connect with other travelers because it seems no one likes small talk, myself included, realizing if we only have a few days together it's going to be real and memorable.

We organize a motorcycle ride and picnic lunch at a nearby waterfall with my Indonesian friends from the barbecue last night. With my hair blowing in the wind, my new pals, and some visions of palm trees, we finally arrive at a small waterfall. No parking lot, no signs, no one else there. We follow the five locals down a hiking path to the waterfall and once there, they all stand with their arms open, bearing huge smiles, as if to say *'Ta-Daaaaa!'*

Our smiles freeze onto our faces as we see litter. Just litter. Everywhere. They want to sit down on the litter to have our picnic. In front of the litter-filled river, with the trash pouring over the waterfall. I can't even see water, only garbage. This is a garbage-fall, not a waterfall.

"Very beautiful?" our host says to us. I can't help but feel a bit slighted, as I remember he'd just called *me* beautiful last night.

NO, it's not beautiful, it's ugly and neglected and destroyed by your country's lack of knowledge about protecting the environment. "Yes," we Americans say in unison, "Just beautiful." We keep stealing glances at each other.

We eat lunch, cracking jokes under our breath about sitting on a blanket over possible cat litter, Jimmy Hoffa's dead body, and the Lost City of Atlantis.

"Well, this is just what we needed to see, maybe we can also teach them about cleaning up after yourself," Joy says.

"It's going to be hard to break a lifelong pattern for them, but if we can make a difference, we have to try..." Jared announces. Finding a perfect plastic bag in the pile, he begins to fill it.

We all pick up our trash while the Indonesians wave their hands saying leave it, leave it. *Oh, right from wrong, insult our hosts or clean the Earth?* We pick up ours and even pick up a bit more in our area since there's a trash can near the road leading in. God only knows what they use it for, but why not show them how we westerners really feel about trash? Maybe they'll think it's cool to copy us and teach their own kids in a similar fashion?

On the way home, Savannah and her motorcycle guy run over a chicken as it's crossing the road. I know, I know, there's a joke in there somewhere, but she has chicken blood splattered on her face. *Gross.* And she's a vegan animal lover who lived on a farm delivering baby horses and calves.

Back home in my room, I get a knock on my door and Mr. Beautiful is standing there. "I. like. you." He says in broken English.

I smile, "I like you, too!" I say. *Awww, so sweet.*

"No, ummm, uhhhh, I *Like* you," he tries again.

Oh, uh oh. "Ma'af jenang gangu saya, saya ada pacar..." I hope I've said it correctly.

He slightly bows, shaking his head as if to apologize and disappears back to his own house. *Yep, I've got a boyfriend, buddy. No Disturbo.*

I head out to use the shower before bed, and I see Savannah on my way in. I tell her what just happened, and she explains how I need to be careful here.

"Women just don't travel without a man in a Muslim country. Ever. They think white women are whores if they aren't with a man," she says.

Plus, it's so hot I've been wearing my tank tops and she says T-shirts would be better, even my jeans, or long sarongs. *Oooops.*

"Thank you! I'd heard there were topless beaches on Bali. I guess I thought they'd be used to tourists?"

"Topless beaches? I doubt that, either way, Bali's a long way from here. I just think most tourists are once-in-awhile day trippers from Singapore and not much else passes through these parts," she sounds concerned. "They're a very proud people. Have you noticed they speak very softly? They think being loud or showing too much emotion is extremely rude and disrespectful. The same with too much skin."

She's right. I need to think about the bigger picture. Would Mr. Beautiful have been so eager to leave me alone if the other Americans weren't here?

I go in to use the facilities and take my first splash-shower. I dip the scooper into the cool refreshing water and pour it over myself. This cold water feels awesome because it's 125 degrees at night with 125% humidity. I wash and condition my hair, shave my legs, and now I ponder. *To trim or not to trim, that is the question.* I'm not dating anyone, so my *'Garden of Eden'* has morphed into the *'Hanging Gardens of Babylon.'* I hesitate only a moment. I trim for myself, dammit, not for any man. Gingerly, in the dim light of the flickering lightbulb, my Lady Godiva evolves back into the Sinead O'Connor of the 90's. It's only after I dry off, get dressed, and put the scooper back, that I see it.

There's a long white worm in the bottom of the barrel.

I've just bathed with a *tapeworm.*

Yep, $2.50 a night in Indo. Not the best ending to a weird, yet hilarious day... *Hey, Jimmy was right!*

Singkep Island

- Meet the Flintstones -
Curtin, Hanna, Barbera

The others are booked on my ferry south to Singkep Island, so we happily travel together. Once we arrive, we all hop onto a large truck carrying hay into town. Jared and I jump into the front seat, the girls in the back. Winding along a steep hillside, dirt road, no guard rail, no seatbelts, we turn a sharp hairpin curve and **BAM,** *collide head-on* with another truck. The drivers get out, *don't say a word to each other*, and pull their mangled bumpers apart, throwing ours into the back with the girls. We all exchange "*No way*" and "*So* *that just happened*" glances. We squeeze past him on this tiny jungle road without another thought about it.

We're dropped off near a group of motorcyclists who then offer to take us to Miss Dita's longhouse. She's a tiny thing, about 65 years old with short black hair, and she speaks great English. She seems somewhat wealthier as it's a proper house, not a thatched-roof hut. Miss Dita rents out rooms for tourists, provides meals, and can play tour guide for us. She has a few empty rooms, but the San Francisco friends insist I share theirs, so we put another mattress on the floor for me.

After the long travel day, we unpack, shower up, and sit down to a wonderful meal. She's made what looks like a salad with peanut butter dressing, called gado-gado. Sounds funky, but it's *delicious.*

Dita wants to know everything about America, about what our hometowns are like, our families, and how we ended up way over here. Jared pulls out a little photo album of his friends and family around California. She loves seeing his family's faces and what our side of the world looks like, probably for the first time ever. And she can't believe we just came through Batam and Bintan.

"They are so dangerous!" she states matter-of-factly.

"Really? Why?" asks Jared.

"Everyone will rob you," Dita answers. "They might stab you, too."

It turns out Dita was born and raised on this island and has never left it, so everywhere else seems hostile and violent.

"What religion are you, Pam? What is your ethnic background?" she turns to me.

Of course, this is the question I'm not quite sure how to answer. I was raised by a hippie mom, and we never once went to church. I would help out in the garden and if there were two butterflies flirting in the breeze, my mom would say "Look! We're the only people on the planet seeing the butterflies right now, it's a gift from God!" My church has been the beach, the mountains, the forest, the desert, the snow.

I attended Sunday School just a couple of times at around five years old when visiting relatives, so I do remember some Bible stories. My biological Dad is half Jewish but converted to Christianity. My stepdad is Jewish, so I was raised with Christmas and Hanukkah, but we never set foot into a Jewish temple either. I've never read the Bible, the Torah, or anything else for that matter.

"Well, I am half English and Christian, part French, Russian, Austrian, Polish, and a bit Jewish," I mention.

"*Jewish!?*" she exclaims.

"Yes, a quarter Jewish," I'm not sure why she picked that part out since I also said five other things.

"You *cannot* say that here. Do NOT say that to anyone else," she puts her finger to her lips to shush me and shakes her head NO, standing to shut the kitchen window.

Really? We're still doing that? This is 1995, Muslims and Jews don't get along by now?

*****

Dita's a wonderful hostess, introducing us to her dear friends, Joyo, a local 20-year-old man who helps with her errands, and Jani, a young business beauty with black hair down to her ankles. They take turns taking us to the market, to soccer games, to the local school, and even invite us to their sing-along group meeting tonight. Tomorrow, the others take a boat to the east while I'm heading west to Sumatra.

We expect to show up at a little old ladies' home for the singalong and have some tea or something. But Dita, Jani, and Joyo drive us out to the middle of the empty jungle. We get out of the car and hike about half a mile through the trees until we come to a clearing, complete with a full *'Gilligan's Island'* stage with a light show, and a full set of bleachers packed with hundreds of islanders.

Well, this is unexpected.

The entire town is honoring us with a show. To our even greater surprise, there's a full karaoke machine with microphones and a DVD movie disk for each song. This is 1995. Music CDs are just getting started, DVDs are a thing in futuristic movies. Each person gets up on stage and belts out American songs, Beatles, Eagles, Creedence Clearwater Revival, oh, how they love their CCR on this island. Each singer is precise, even serious, taking full bows and curtsies after they're done to rounds of applause with standing ovations from the audience.

For the finale, they insist we sing a song in return. We don't know what to do. All of us are the exact opposite of performers, yet we don't want to insult our hostess, or apparently the entire island. We go up on stage, full spotlights in our faces, and we've agreed on a popular song. The music begins and we belt it into the microphone.

"Flint-stones! Meet the Flint-stones! They're the modern stone-age fam-i-ly! From the—Town of Bedrock—They're a page right out of His-tor-yyyyyyy!"

When we're done, there are *crickets* in the audience. No one is applauding. This is *awkward*. I think we *did* insult them by our light-hearted choice. In our own country karaoke is a silly hobby, not a true live performance at the Sydney Opera House. Someone thankfully begins the slow movie clap, others join in, and we dash off the stage with beet-red faces.

We're all giggling as we head back to the house. Dita wants to make sure the Flintstones song is a good song in the USA. "Of course!" we all reassure her. "Everyone in America knows that song." We hug her and thank her for this magical night as we head off to bed.

The next morning, we swap addresses and goodbyes, and the San Franciscans are off to their next adventures. My boat's late, so Joyo offers to wait with me. Hours later, baked from the hot sun, we're still waiting. We joke around. I teach him California slang. "Stoked," meaning "happy," is his favorite. It's Monday evening by now and Joyo has just discovered my boat isn't coming at all. Cancelled. But on Thursday a cargo boat is coming and heading to Sumatra, a 17-hour boat ride. "I'll take it!" I say enthusiastically, not grasping the full meaning of *cargo boat*.

I am, however, understanding the meaning of "Indonesian time." Basically, there are zero deadlines, and things get there when they get there. But really, I'm trapped on a tropical island? With new friends, a roof over my head, and good food? This is just fine.

Joyo invites me to his home to meet his 104-year-old grandfather. He's lovely, and while we are sipping our tea, I notice their neighbor has kittens a few weeks old. I choose one calico kitty, with his face half black, half orange, like he's wearing a Phantom of the Opera mask. A gift for Dita because she's been upset about a mouse in her house. I bring the kitten in and she has tears in her eyes she's so thankful.

"I name him Zorro!" she says with gusto, clapping.

"That's purrrrfect!" we laugh. "Hmmm, can you afford cat food?" I have the thought, I mean, really, I just brought her a new pet, without asking if it's okay first.

"What is this 'cat food'?" she wonders.

I think of a new way to say it, "I mean, what are you going to feed Zorro?"

She says, "Oh, just a bit of whatever I'm eating for dinner, you know, fish, rice, things of this nature."

Of course. Everyone in the world fed their pets before America came along and marketed cat food.

"Hooray for Zorro!" She shouts with glee and shows him the mouse holes in her walls.

The next day Jani's having breakfast with us. They remind me of the custom of not touching things with my left hand, as I reach for my glass of juice. I remind them it's not our custom and I meant no disrespect at all, yet can I ask a few questions? I genuinely ask how do you butter toast if you can't touch it with both hands? Both women look embarrassed, quickly busying themselves hoping they don't have to answer. *Ohhh, you DO touch it, but no one talks about it. Eeeewww.*

Jani clears her throat. She asks me something quickly in Indonesian. We stare at each other a moment as I realize I'm supposed to answer, and she realizes she didn't speak English. Dita spits her tea back into the cup, she's laughing so hard. It turns out, Jani's relatives are getting married, and I'm invited to a wedding.

"Really?" I ask, "They don't even know me. Are you sure there's enough room for one more?"

"Of course! You are our guest," she replies, smiling warmly.

"What should I bring as a wedding gift?" I ask, as that's our custom in the west.

"Nothing at all, just come with us this afternoon," she's excited for me to meet her family.

We ladies get all gussied up. I whip out my awesome black sarong and black T-shirt with my sandals. Hey, I'm backpacking. But I also still have mascara and a one-dollar eyeliner in a flattering cinnamon raisin color that has become my go-to for everything. It's my eyeliner, my eyeshadow and after smearing a bit on a fingertip and rubbing on my face as blusher, and my lip-liner that when I fill in my entire lip and pair with Carmex, looks like a real lipstick. I have a barrette to pull back my uncut mop in a half-up and half-down 'do. Ready in five minutes.

Jani, Dita, Joyo, and I walk a few houses down the street and bada-boom, bada-bing, we've arrived. Long tables with chairs line the garden, food everywhere, like the ultimate Thanksgiving Dinner. The area's decorated with strings attaching those Styrofoam white shipping peanuts, hung like streamers, across the patio and throughout the trees. Pink striped plastic grocery bags were blown up, tied in knots like balloons, and hung up around tree branches. Everyone's dressed up. The men in button-down shirts with sarongs and flip-flops, the women in long skirts, blouses, and fancy hair wraps. I'm introduced to everyone and greeted warmly.

We sit down to eat and I'm wondering if we're late, because in the USA, the reception is always *after* the nuptials. Hours later, after cake's been served, Jani asks if I would like to meet the bride and groom.

"Absolutely!" I was wondering which couple they were this entire time.

We enter the house together, just the two of us, as another person's heading out. Jani explains the couple sit inside the home all day with just the Muslim Imam and it's an all-day ceremony. Only one person at a time can meet with them. One by one, people wish them well, so as not to take the focus off exchanging their vows before God. She points me through a hallway and waits for me while I must go on, by myself, through a stranger's home, to meet the bride and groom of a wedding I've crashed. *I hope they heard I was actually invited?* Yikes, now I'm nervous.

I turn a corner and see the couple and the Muslim cleric. The groom is dressed in a fancy Indonesian suit with a fez-type hat and a sash across his chest, while the bride looks glorious. She's wearing a full gold tin-foil Cinderella gown. Glamorous full makeup, with hair swept up into a gold jeweled crown. She wears huge gold earrings, bracelets, and necklaces piled on like Cleopatra herself.

I smile and they just stare at me. I tell them softly how grateful I am to have been invited to their wedding, to share in this beautiful moment, offering my warmest congratulations. They stare at me and say *nothing.*

Hmmm, maybe they don't speak English? I bow deeply. Nothing. I gesture to my camera asking if I can take a photo. The Imam nods and they say nothing.

I'm waiting for them to smile and they're waiting for me to take the picture, not smiling. I take one, hoping the clicking sound will jolt them, reminding them to smile, then I can take another with cheerful looks on their faces.

"Okay, say 'Cheese!'" I try again. They continue to stare at me. Not smiling. *Are they angry that I'm here? Maybe she's pregnant and this is a shotgun wedding? Maybe it's an arranged marriage and they just met five minutes before I arrived?* I want to shrink down and die. Or maybe I should dance the Charleston while singing, *"Hello, my honey... Hello, my baby... Hello, my ragtime gal!"*

I put my hands together in a slight bow and start to back out when the officiant asks me to wait. We all just stare at each other for a few minutes. Me smiling, them not smiling, as usual. Finally, the groom, saying nothing, reaches down into the basket on the floor next to him and hands me a muffin and an egg. No plate, no muffin paper, just into the palms of my hands. I don't know what exactly is expected of me. *Do I eat them now?* I'm stuffed to the gills from the reception before the wedding. And the egg looks and smells really *off*. Plus, he handed me the muffin from his hand, and I still can't get over the no-toilet-paper-use-your-hand thing.

This is weird and hilarious.

"*Teri makasi*, thank you," I say and bow again slightly, turn, and run out, with my muffin and my egg.

I don't want to insult the family by asking the wrong questions, but *oh, I have so many questions.*

I tell Jani what happened and what I said to the couple. She thought it was lovely and I did the right thing by not eating the gifts yet. I'm supposed to wait 24 hours and then eat, for some reason. I tell her I tried to take a picture of them, one they agreed I could take, but that they didn't smile. She explains this is a very serious day. They're not allowed to speak with *anyone* all day but the Imam. In this way they can really show God they're serious about their promises to each other. *Good thing I didn't try to make them laugh, they're like Buckingham Palace guards.*

"It's a day for celebration, but not getting drunk and throwing a garter at friends. We find this style in America offensive to making vows before God." I kind of like taking the vows more seriously. Americans could learn from this, not just "Hey, I'll give marriage a try, but when I'm sick of picking up his dirty socks from the floor or she gains fifteen pounds after having my five kids, I'll just get divorced."

So finally, I ask about the egg and the muffin.

"The egg is a blessing for health, especially fertility, that they will have a large family, plenty of children, and wish the same for you. The muffin is a blessing that there is always enough prosperity, food, and comforts that will always prevail, and they wish the same for you," she tells me, beaming proudly.

"That is lovely!" I'm excited to receive such gifts. Man, I wish I'd had something to give them as a present in return. "So, everyone gets this when they go say their blessings?"

"No. Only *one person* gets this gift at each wedding," she says, delighted they chose me.

I'm truly touched. And I'd just come for the free food. *Am I an asshole or what?*

We dance the night away in the garden and I'm still clutching the egg and the muffin the entire night. Back at Dita's, I look again at my gifts and realize I'm *never* going to eat these and roll them up in paper towels. I gently place them in the trash, so Dita and Jani can't see what I've done. The next morning, I tell them how yummy my gifts were.

Now I'm an asshole.

*****

My cargo boat has arrived. My pals take me to the harbor for my ticket. It's a dollar, much cheaper than I thought it would be. And when I see my vessel, I understand why. I'm reminded of a bigger '*Buy-A-Newie.*'

I'm going to be on *this* for 17 hours? First in the ocean, then 12 hours down the Sungai Kua and Sungai Batanghari Rivers to Jambi, a large city on the island of Sumatra. *If it makes it.* I'm the only tourist, the only westerner, and definitely the only girl. There must be 40 men on board and everyone's just lounging around on the decks—and by lounging, I don't mean sitting in a chaise lounge by the pool with a cocktail. Everyone's lying on the wooden floors, and only a few have the luxury of a flat foam square. Zero chairs.

After hugging my friends, with Dita warning me that Sumatra can be very dangerous, I climb aboard and find a spot on the floor. Good thing I'm wearing my jeans for travel, but it's already sweltering. I have a Walkman and my cassette tapes with me, but it seems disrespectful to whip out something that no one else has. Plus, it just waves wealth to anyone who might want to rob me.

Everyone's staring. I just smile and close my eyes and pretend to sleep, head on my pack. Moments later someone taps my foot. I open my eyes and a man is gesturing for me to take his piece of foam. How sweet. I smile, grab it, and pretend to go back to sleep. I'm sure I really do fall asleep for a while due to the sway of the ship in the ocean.

Hours later, I open my eyes and everyone's still staring at me. I roll over so I'm facing the wall and I see a cockroach run by my face. It's so large, about the size of my entire hand, fingers included. I'm reminded of a joke from a female comedian, "'*A bug is on my back, could you just knock it off me?' 'Just a June bug, won't hurt you' 'It has a* dog *in its mouth*'!" I reason with myself that if I hadn't opened my eyes, it would've just crawled right on by. Maybe it's been going back and forth for hours.

Once we hit the river, the sailing slows way down to a putt-putt. Thick jungle leads right up to the river's edge as we pass small towns... and by towns, I mean a few huts in a row. I feel like '*Rambo*' going down the Vietnam rivers in search of missing soldiers. Locals come out in their canoes filled with things for sale like sodas, bananas, and fish. I get some funny looks, which reminds me how rare this is for a girl like me to be on this type of excursion.

Some of the men onboard approach with a soda they've bought for me, gesturing for me to join them in a meal of rice and fish. I'm starving, so I agree, feeling a bit bad that I'd pretended to sleep around all these nice people. It hasn't escaped my attention that I could buy this entire boat with the money I have on me right now, and these gentlemen who probably make $100 a year, are buying *me* food. I'm truly humbled.

We all sit cross-legged together and reach into the, yes—*wait for it*—White Rice, with our fingers as scoopers. They slice down the side of a massive fish, lift the head, and pull out the entire set of bones in one swift motion. They show me how to peel back the skin and the cooked fish is perfect. We feast. One man begins to practice his English slowly with me. He wants to know about America, where my husband is, and how much money I make as a waitress in California.

I lie. "$300 a month."

He turns in shock to tell his friends. "You much rich!" he tells me.

Jeez, I'm glad I lied. Do they want to rob me now, or should I buy them a soda at the next stop? I wonder what the reaction would've been if I told them the truth, about $2,500 with tips.

I tell them I don't get to keep most of it, because it goes for income taxes, sales taxes, and rent. I already know there's no sales tax in Indonesia, and most people's homes are passed down through the generations.

He grabs the fish head that's between us on the floor, and with his fingers, proceeds to poke its golf-ball-sized eyeballs out of its fish face and offer me one to eat. I can feel my stomach turn at the thought.

I look at the eye. It's slimy and bloody... I look at his smiling, glowing, giving face... I look again at the eye. *It's looking at me, so, ummmm, no thank you.* He shrugs and pops it into his mouth, squirting eyeball juices down his chin.

Yes, I'm representing America. But I ***cannot*** *eat fish eyeball. I cannot on a boat, I will not in a moat, I could not, would not, on a ship, eat that fish eye and watch it drip. I do not like it, Pam, I am.*

The next couple of towns we pass still have the people in canoes selling things, but my new friend says now they're asking about the white woman on board. With news spreading fast by motorbikes between villages, it seems my reputation has preceded me. They all want a look at me as I'm smiling and waving to the gathering crowds. People have piled into small canoes with their kids, who may never have seen a white person before. I'm enjoying every minute, but it also feels like I'm a zoo animal, so I'm left wondering if this is what it feels like to be a famous person back in LA.

After the meal, the sun sets and we settle down for the night. But it's so flippin' hot it's become unbearable. There's a quarter of an inch between each of the wooden slats on the floor where I can look down to see the motor. The heat from the engine is coming through those slats and I'm being steamed like vegetables on a grill.

Many of the men have gone on top of the boat for this reason, so I climb up. A *teensy* bit cooler, and just half of the engine steam make this a fantastic spot to sleep under the Sumatran sky.

The night is so crystal clear I can see billions upon billions of stars, even a shooting star. Gazing upon the raw, profound beauty that is the universe, I realize no one in the world knows where I am this very moment. I feel insignificant. Tiny.

I visualize the vast ocean, how many islands and countries come between me and my home in California. How many people, places, and things just go on about their daily existences, no matter if I'm here to witness it or not. This cargo boat would be chugging along this river even if I'd never taken this journey. The waves would still be crashing against all the shores around the globe whether or not I'm there, like they have for many millennia.

I wish I'd been born centuries earlier and been a proper explorer. The absolute courage of the Vikings, Columbus, James Cook, Magellan.

At first light we pass kids playing in the water, women washing their family's clothing against the rocks, and large oxen along the river's edge. The jungle is beginning to thin as the city springs up. More homes line the river now. I buy a quick egg breakfast for my new friends and we arrive into Jambi.

Sumatra

Jambi

- People Are Strange -
The Doors

I walk onto the docks and everyone's staring at me. I head into Jambi and everyone's staring at me. I duck into a crowded food court and order something with a picture in the menu. Everyone's still staring at me. I catch their eyes and smile, hoping they'll realize they've been caught, and they should look away shyly, like the polite game we do back home. Nope, they just smile back and keep on staring. I guess it's not rude here.

Lots of people look like regular Indonesians, but when they open their mouths to smile, I catch a glimpse of their teeth and recoil. *Jet black teeth.* I've never seen anything like this and am wondering *if they drank the tapeworm water? Is their toothpaste made of charcoal? Is this like when I drink too much red wine and my teeth go slightly burgundy until I brush?* So many people have it, I'm just trying to hold a steady smile and act like it's no big deal. Picking one's nose and spitting onto the ground doesn't seem to be out of order here either, even in a restaurant. One man has his bare foot up on the table and is clipping his toenails. Toenail shrapnel goes flying onto the full plates the waiter's carrying. *No biggie. All good here.*

I find a small hotel for $2, and after three showers in a row, sleep the rest of the afternoon, through the night. In the morning, loads of kids follow me around, yelling "Hello Mister!" giggling and sneaking up to pinch hair off my arms. Local's arms are smooth as a baby's butt, so they're curious about the way I have a bit of body hair. It's painfully funny.

A lone shaggy dog sits next to me on the street. His eyeball's hanging out of its socket, he's covered with ticks and scabs. I cringe and don't pet him. My heart breaks as he forgives me and walks away. Truly my deepest travel regret—how I'd selfishly avoided giving him a moment's joy.

Hmmmm. I could head north to see the orangutans at Lake Toba, see the surfing spot Nias Island, where beach huts are 10 cents a night, or go south to the beaches and ride elephants through the jungle. *Yep, Dumbo for me.* I buy the ticket south with a beach stop first. Hours upon hours on a slow, non-air-conditioned bus with lots of locals smashed on board, with their goats and chickens.

I wish I could share this.

I practice speaking Indonesian with my seat mate, who doesn't really speak English. Maybe he knows a phrase or two, but we end up making small talk. We share the same stop, plus some motorcycle rides to bungalows on the beach. It looks like my place on Bintan Island. A mattress, a tin door, and *that bathroom*. He stays next door. It's so flippin' hot. There's no AC, or even a fan, so it's beyond hotter than the sun, even at past midnight now. I sit out on the porch and my friend's still up, too. More small talk. He's lovely, we say goodnight and he's gone before I'm awake.

My breakfast is tasty while I gaze out to the beach alone on Sumatra. I've learned so much of the language and the people here are kind, generous, warm, and funny. I lie on the beach in my bikini since there's no one around, read my book, and fall asleep in the shade of the palm leaves.

A noise jolts me awake. *British troops?* There are now 30 Sumatran people sitting in a circle around me. Families, mothers, grandparents, children. *God, how long was I sleeping?* The rest of the beach is empty, and these folks seem to be having a picnic lunch on my face. They're literally touching me, my hair, my toes, while laughing to their family and friends that I have polish on my toenails. The children giggle and call me "Spaghetti Hair!" I smile and gather my things. They don't even seem to think they are a tad intrusive. Maybe I'm disrespectful wearing my bikini?

Another long, melting, sweltering, sauna night ahead. I look up on my thatched ceiling and there's a huge tarantula. *Not joking*. I run out and, even though I know my talking buddy's gone, I knock on his old door and a new Indonesian man answers. He literally rubs his eyes as if he cannot believe there's a young woman knocking on his door. I smile, waving to gently say "*come with me...*"

His face lights up as if this is a fantasy he's not going to pass up. The porno pizza delivery guy kind of thing. He leaps out the door, follows me closely to my room, combing his hair into place, tucking in his shirt. Once inside my room, he gets a look on his face like Joey from '*Friends*' that says, "*How you doin'?*" I point to the spider on the ceiling. His face goes from "*Yippie!*" to "*Oh, man!*" in a flash of a second. He's a gentleman, removing the offensive creature as I thank him and shut the door.

A beach stop wasn't what I needed and it's time to go. At the crack of dawn, I buy a bus ticket to the Elephant Reserve. The bus leaves in one hour, so I rush back to pack, when there's a knock at my door and it's my old bus talking buddy. He's back. He aggressively grabs my hand, walks me to some friends of his, who have obviously been invited, as if to say "See, that story of my western girlfriend is true!"

I'm instantly annoyed. He leans in to kiss me in front of his friends, so I pull away, scowling, and walk back to grab my bag. *How rude*. He was such a gentleman the other day, never even tried to touch me and now I owe him something? If I owe anyone anything, it's the exterminator dude, and it's like $5, not a kiss of any sort.

I hear *Metallica* blasting loudly. I turn in time to see my minibus pull up. *Sweet! My kind of music*. I grab a seat among 10 Muslim teenaged girls. They're staring at me with their hair covered. On the ceiling's a huge poster of *Iron Maiden*, with *Eddy*, their skeleton mascot ripping through the car. *My very first tape I'd bought in the 7th grade. Awww*. It feels out of place, even funny, at 7:00 a.m. with very religious young girls. A hilarious start to a better day. I'm amused as I realize I'm on the local school bus.

I switch buses a few times, and as I'm transferring, some locals gather around me. It's quite normal as they'd like to practice their English, have their children meet the Spaghetti-haired traveler, or try to see if I need assistance getting where I need to go. I'm pointing at the map to the elephant reserve, when a man grabs my butt so lightly, I'm not sure what to do. Now *that's* unusual. I know in this country men do NOT touch women. Even their wives, not in public. It's quite common to see men holding hands and walking arm in arm even though they're not gay.

Everyone scatters.

I reach into my back pocket for my $10 and yep, it's gone, baby, gone. I've been pick-pocketed. Oh well, Jay had taught me to keep a few bucks handy just for pickpockets. They run away feeling as if they scored when, in reality, I still have over $2,000 on me.

Telukbetung

- Run Through the Jungle -
Creedence Clearwater Revival

I arrive at the Satwa Elephant Lodge. I'm the only person here besides the guide. My room's like a regular hotel room, with no AC but a table fan and a ceiling fan. Dark hard wood floors and dark wood furniture seem extravagantly luxurious after the places I'd been staying. There's even a regular western toilet and shower down the hall. My double bed has a mosquito net and there's a mirror above the dresser.

Ugh, I haven't seen my own face in a week. The last place was Dita's house, a lifetime ago. I need to tweeze my eyebrows before I go ride through the jungle on an elephant. *Seriously, pictures.*

I'd picked up malaria tablets before leaving the USA, along with shots for yellow fever, cholera, typhoid, hepatitis A and B, meningitis, and tetanus, along with boosters for polio and measles. I hadn't been 100 percent sure if I would continue north through Africa, so better to be prepared. My grandfather had contracted malaria living in Africa and it didn't sound fun. Basically, a heavy flu that lasts two to three weeks and returns each year for the rest of your life. You may or may not die each time. *No thanks.* As I swallow my malaria tablet, I'm so glad I did take those precautions. I'd just begun taking the medications upon landing in Kuala Lumpur.

My guide, Rimbo, sees me taking the pills.

"Malaria?" he points to the tablets.

"Yes, just in case," I say with a shrug and a smile.

He's quite friendly and says in *Yoda* English, "No, no, no, tea we make you." He gestures for me to follow him through the jungle for a bit before we see the elephants. It's still early in the morning, so I'm again filled with the flow of adventure. I'm wearing my mother's yellow embroidered tunic shirt, long jeans with my boots, with my hair in Indian braids. I'm ready for today.

I follow Rimbo into the thick rainforest. Stepping over the first tree roots, he shouts "Ahhhh! Snake!" I know it's a joke even before he starts laughing. Some people give off a warm and witty vibe, like this guy. I crack a joke back and pretend to nudge him in front of me so he would get the snake bite instead of me. We bond.

The rest of the hike is like this for the next hour or so with just the two of us and his machete in the most thick, dense jungle I've ever seen. It's stunningly searing, humid, and dark, the canopy blocking out the sunlight. Feels like we're in a sauna after a gym workout. The jungle's a symphony, alive with cicadas, exotic birds, and monkeys. He teaches me about the flowers and plants, picking them as we go, for my own magic malaria prevention potion.

As we get back to the hotel grounds, Rimbo stops me and kneels to look at my boots.

"Leeches, yes," he says calmly, like he's saying it might rain later.

"Leeches?" I wonder if he's joking again. "What do you mean 'leeches'?" *Please be lychees.*

"Leeches... you shoes," he points. "Feel warm from first person walk in jungle, fall from tree, second person. They go lady bits. You check. Here lighter, burn. Here balm for stop bleeding."

"Seriously? Not joking?" *Please be joking.*

He smiles and motions for me to hurry. *Fantastic, not joking.* I take the tools for further inspection.

In the bathroom I strip down. There are two more wrapped in my shoelaces and—*gulp*—three black squigglers *crawling up my left leg.* They're little black sluggish creatures, leaving a dot of blood where they'd latched on, until climbing north again. They secrete something weird that doesn't let the wound heal or the blood coagulate. They get fatter, like a water balloon, filling up on my blood. One above my ankle, another mid-calf, and the last behind my knee. *SO VILE.* I feel like the boys in the movie '*Stand by Me*' when they get leeches all over them in the river, even one on a poor kid's nuts.

I can't feel them at all, really. Zero pain. Still, I grab my lighter and sizzle that first little fucker. He curls up as he's un-suctioning, and in a dramatic slow motion, falls backwards so I stomp him with my boot, complete with a miniature gore explosion. On to the next one, and the last. I quadruple check my "lady bits" and the rest of myself and yep, I'd found all the little bastards. Okay, dressed again and ready for some malaria tea before I hit the elephant jungle trek.

The tea tastes like a dirty hippie after three days at Woodstock, but what the heck, I made it myself from the plants I picked along my leech-filled jungle hike. It's awesome. Along with 'jaffle' snacks. I order it because it seems like a misspelled 'waffle.' I'm wrong, but in a delicious way. They're pressed sandwiches like a hot pocket, but filled with sweet, like banana and honey, or savory, like a fried egg— with a side of rice—but *yum.*

Now, I'm ready to meet the elephants.

I have a sense this won't be scary at all, like getting chased, I'll get to see their gentle sides. My palms are sweaty and I'm giggly, I can't wait. We head over to the Way Kambas National Reserve and I see about 10 hanging around a bamboo hut stable. *And there's a baby one.* I go right up to the baby and he leans into me while I pet him. I feel his fuzzy little hair on his toddler elephant head. He's dancing back and forth, and he keeps wrapping his baby little trunk around my arm, he's trying to pull at my braids and feel my face.

I love him.

His mama's nearby and I give her some treats like Rimbo instructs me to. She knows I'm a friend, and she doesn't seem worried in the first place, which shows me they must be treated fairly here.

I have to say goodbye to my little darling pal as I'm brought to my new and much larger pal I'll be riding. Hmmm, these elephants are slightly different from their African cousins. Still huge, they're shorter here, with much larger foreheads and smaller ears.

My trail guide climbs onto the pachyderm, one leg behind each ear. He motions for me to get on behind him. He's made it look so easy. I stand there for a moment, studying the elephant's body, not sure how to proceed. Gracefully, the elephant kneels its front legs to allow me to climb up. I step up onto his bent knee, and he even holds his trunk to my hand to help steady me. I grab his ear and swing my other leg over the back of his neck, behind the guide, like riding a motorcycle together.

The scent of the animal, the thick skin, and the heat from his body, all make it *real*. I'm alive in this moment. I can pet the course, wiry hair on his head and neck easily. It reminds me of the bracelet made from elephant hair that my mother had gifted me from her African trip. Mine had been in my stolen moving truck, never to be seen again. The thieves wouldn't have a clue the sentimental value of something like this. I'd bet they couldn't have even known what it was, it just looked like a black wire bracelet that a little girl might make at summer camp. It probably ended up in the trash. I secretly wonder if I can pull out a few hairs from this elephant, if he would even notice, or if he would grab me around the waist with his trunk and toss me down into his gargantuan dung heap.

Surrounded by trees, vines, and jungle noises, my guide taps the creature on his temple with a little metal hammer. *Hmmm, I surely don't like that.* He gives a little shout and we start moving. I'm riding my first elephant through the jungle of Sumatra. We sway back and forth, back and forth, taking a well-worn path into the dense rainforest. The bones in his shoulders move beneath our butts. I'm giggling into my guide's ear, causing him to get the giggles as well.

Our elephant is gently holding back vines and pushing bushes aside with his trunk, and my guide is tapping on the left or right temples and giving short commands. One big shout from us and our big daddy lifts his trunk to trumpet, pushing down a small tree like he's snapping a toothpick. Because I'd ridden horses quite a few times in my life, I keep subconsciously kicking my heels in to make him go, only he's gigantic and my heels are resting on his upper front legs, not his underbelly.

My friend keeps pointing out exotic birds, lizards, and overhead monkeys. Pure joy. He hands me the hammer, jumps down, and walks next to us. I slide forward a little on my very own mammoth. I don't have the heart to use the hammer, so I gently tap his temple with just the palm of my hand, and it works. We wander through a river, the guide up to his chest in the water. With a short command, my elephant sucks up some water and squirts it over my head, just the right amount, misting me perfectly on this scorching day.

After riding in a large circle for over an hour, we break for lunch for us and the animals. Back at camp, my elephant again kneels and offers me his trunk to balance myself as I step down. It's slimy as my thumb accidentally penetrates his nostril. My guide and I are all smiles, and I buy a large bag of peanuts to feed the herd. It seems they're taking turns, waiting in line for a peanut. The baby one comes running over, pushing and shoving until it's his turn, just like a three-year-old boy in line for cotton candy at a carnival. The others get impatient, and one even tries to pull him back to his spot in line. *Manners first.*

Once the bag is empty, as Rimbo and I begin to walk away, the baby elephant runs over at a full gallop. I freeze up, not quite sure what's happening. *African stampede?*

"He like you," Rimbo says. "He say 'Bye friend'."

This darling toddler elephant keeps wrapping his trunk around me over and over, leaning into me, like an *adorable hug.* I stay another 30 minutes playing with him, then sadly, we must go.

"When return, he remember you," Rimbo reminds me. "20, 30 year more, he remember you."

I can't help but tear up a bit, wondering if I'll ever be so lucky as to return one day. Maybe with a husband and little children of my own? How fascinating, to show this elephant *my babies* one day.

Back at the hotel, Rimbo and I are still alone. He cooks me a great dinner, white rice of course, chicken, veggies, and some egg all mixed with a type of soy sauce. We're still cracking jokes. When it's time for bed, for the first time, I'm a bit afraid of being in the jungle. At night. All alone in the hotel, as Rimbo's got his own living quarters somewhere nearby, just off the property.

The sun has set and the generator's turning off in minutes, so I light my lantern, tuck in my mosquito net, and read a book. I'm really enjoying '*The Imajica*' by Clive Barker, about a traveling person going between different worlds. I hear some funky jungle noises and decide to sleep with the lantern on so I can see if there's anything weird sneaking into my room. This entire trip I've had a paring knife with a small cap on it. I grab it out of my bag and stick it under my pillow. Just in case.

I wake up around 2:00 a.m. with the need to use the loo. I uncrack my eyes. I'm so into my book, I'd fallen asleep with it in my hand. The lantern glow is still on high, so I see my entire mosquito net is *absolutely covered with enormous jungle insects.*

There are huge praying mantises as big as my forearm, titan beetles the size of my shoe, katydids and rhinoceros bugs, firefly larvae, spiders *(yikes!)*, cicadas, millipedes, mosquitoes, moths, butterflies, and countless smaller bugs. Thousands of them swarming in my room, all over my floor and on my mosquito net. Along with bugs that like to eat smaller bugs.

My biggest nightmare. And I still have to pee.

Oh My God. What am I going to do? I'd thought the dark wood furniture and flooring was beautiful this morning, but now it's camouflage for the dark creepy-crawlies. Thank God I have the light on or else I wouldn't have seen all of this. You'd think Rimbo would've warned me about this. It must happen every night. *Should I just pee the bed? Can I reach my cup of water and go in that?* I quickly grasp that I cannot do either one. What if the bugs enjoy the scent and find a way into my protected bed?

I scan the mosquito net for any small holes. So far, there are none that I can see. I remember back to the movie '*Indiana Jones 2–The Temple of Doom.*' The actress, Kate Capshaw, walked on a darkened crunchy path and when she lit a candle, the crunchy sounds were huge bugs. They crawled all over her, in her hair, down her nightgown. I'd read she'd been so terrified that the only way she would do the scene was if Steven Spielberg himself stood just off camera and had bugs crawl on him too. In real life, he married her.

Okay, deep breath. I'm going to make a run for the bathroom.

In this day alone, I've been manhandled at the beach, stunned by the heavy metal school bus, been pick-pocketed, did a jungle trek, burned leeches off my leg, made rainforest medicine, ridden an elephant through the jungle, wrestled with a baby elephant, and now awaken to the '*Temple of Doom*' scene.

Just your average Tuesday.

My Grandpa would be proud.

I pull my hair into a tight bun, but leave some hair covering my ears, pull the neck of my shirt up onto my head like the cartoon '*Beavis*' and I tuck my shirt into my sarong. I can see my flip-flops just near my bed, with only a few bugs in them. I reach my arm out, shake them off, and bring my shoes into my bed, quickly tucking the mozzy netting back into place. I survey the damage. Some bugs shifted slightly from the moving net but other than that, most stayed in place. I slide the flip-flops on and take another deep breath.

I make an opening just large enough to slide out my legs, then hips, my head, with my arms last, so I can re-tuck and run. I'm bombarded by the flying bugs and proud of myself for taking all these precautions. I open the door to my room and head into the darkened hallway to the bathroom a few doors down. I keep swiping my arms and legs to make sure I'm all clear and quickly use the facilities, realizing as I bare my ass and lady bits yet again today, it's the perfect time for penis envy.

Stepping back into the dark hallway, I can see my brightly lit room. It looks like I'm a bug scientist and they've all escaped in the apocalypse. As I run into my room of mayhem and shut the door, it dawns on me, wisely, that all the bugs are *only in my room*. Not in the hallway or the bathroom.

Because I left the light on.

I slide into my bed, still swiping my limbs, sliding in and re-tucking realizing that a 'bug scientist' I'm definitely *not*. Everyone knows if you leave the lantern on, all the bugs would be drawn to it. I mean *All of the Bugs*, even a few from nearby Borneo and Sulawesi. I'm sure my Vegas-light was just a beacon for one huge insect rave party.

Well, I sure can't turn it off now.

So, I just sit and watch the bugs from inside my safe netting until the sun rises a couple of hours later. I must've passed out, because next thing I know, I hear knocking on my door. Rimbo wants to know if I'd like some breakfast.

"Yes, please!" I croak out, still terrified. I tell him I'll be right there and slowly check out my situation. No bugs on the net. No bugs flying. No bugs on the floor. *Did I dream this? Hell no.* But everything looks so... normal. I slide out and when I tell him how my night went, he can't stop laughing.

"Bugs like light! Light no good!" he squeaks, barreled over with tears running down his cheeks. We both laugh at me throughout breakfast. *Yep, I'm here for your entertainment.*

I'm not quite sure what to do next. I'm just a few days away from meeting Batam Jimmy in Jakarta. I should get a bus, then ferry, then bus, over to the city on the next island of Java. Just as I'm looking at the map, wondering how to proceed, I hear a car drive up. There are two ladies in the seats with brown curly hair. 'Spaghetti hairs' like me. It's a tourist woman and her teenaged daughter climbing out of the car, with a local driver.

They're speaking French to each other and English to the driver. I give a smile and a wave as I pop open the screen door, thinking my usual sarcastic thoughts. *Good thing there's a screen on the front door, you know, to keep the bugs out...*

The women turn and see me, smiling and waving as well. They come in and have some tea with me and we exchange life stories. There's a school break in France. The mom, Mimi, wanted an adventure with her daughter, Cozette, who's about to start her last year of high school and is deciding which university to attend. They'd hired their driver and he's been taking them around Sumatra for a week. They must return to Java tonight to catch their flight out of Jakarta the next morning.

"Wow, I'm just heading in that exact same direction," I smile.

"Why don't you join us in the car?" Mimi asks me. She turns to the driver, "No problem, right?"

"*Tidak apa-apa,*" is the response. Great, no problem!

"There's not much space, but if you don't mind, you can stretch out on top of our luggage in the back..." she continues.

"Sounds fantastic!" I can't believe my luck.

"We are just going to see the elephants for about an hour, come back here for lunch, and then we'll be off!" she says with a smile.

This is perfect. New friends, a free ride, and some guidance in the big city. I decide to shower up and pack my bags while they see the elephants, so I don't hold them up. Once they return, we enjoy a leisurely lunch and enjoy some great conversations.

"So how did you end up in Sumatra of all places?" I ask.

It turns out Cozette wanted to see the orangutans in Lake Toba to the north and it might even be what she studies in college. Maybe join the team either here, or in Borneo, after her studies.

"And how about you?" Cozette questions.

I tell them all about my moving truck getting stolen in San Diego, meeting J.J. at the ski resort, and traveling ever since, just having adventures.

"Do you mean *after* your studies? You went to Uni first, *right*?" Mimi has a pleading look on her face. *Oh.* Cozette doesn't really want to go to school anymore and she needs a bit of help from me. I guess I'm a—*gulp*—role model. *Oh no.*

"Of course," I say, thinking this will be the end of it.

"What was your degree?" questions Cozette.

Ummm, nothing. Waitressing, concerts, beer, and sun tanning at the beach. I think quickly of how proud I am of my sister, Sandy, who graduated from high school with honors in math and went on to put herself through UCLA with a math degree while working full time. She took Calculus in the ninth grade and went nine years past that. In math. *Math.*

I steal her life for the first time.

"Really?" Cozette beams with excitement. "Did you take this *!@#$%^ class and that *&^%(@ class?"

"Of course." Now I'm panicking and don't even know what she's actually said. Looks like she's a math genius as well. Fantastic—I *stink* at math.

Three years of algebra and I'd passed with a 'C' or worse. Once you add the alphabet into a math equation, I think everyone can agree that Satan is now involved.

My only year of college was with my-sister-from-another-mister, Katie. We'd taken every single class together throughout all of high school and now in college, we'd done the same thing. The first thing the teacher said was, "If there's anyone who doesn't belong in Combinatorics with Statistics in Actuarial Applied Mathematics, you need to exit now." Katie and I sat there with our noses in the books, blushing, trying not to laugh through the entire hour. We were *so* dumb we couldn't even find the right math *classroom.*

I'm quickly reminded that my usual brutally honest self is much easier to be. I guess I've told a few fibs on this journey to feel safer, but really, it's too much work. I'm just too lazy to keep a lie going. Now look what I've done. I already can't wait to get away from them before I'm caught in my own web of deceit and we're going to share a ride together for eight or nine hours. Oh, joy. I quickly change the subject and go grab my backpack.

"Where are the rest of your things?" Mimi wants to know.

"This is all I've brought," I say.

"You've been gone for how long now?" Cozette asks, surprised.

"Almost five months now," I say with pride.

They quickly turn to each other and speak French rapidly. Whew! They care more about my backpack than my 'faux' degree. *See, I speak a little French.* With a huge hug to Rimbo, and his warning that Java is "much more dangerous than Sumatra!" I climb into the car, on top of their eleven bulky suitcases, lying face up with my nose grazing the car ceiling, and off we go.

As I wake from a nice long sleep to soft French music, we drive straight onto the ferry. Onboard, we go to the deck to stretch our legs for the few hours out to sea. The driver teaches us about the Krakatoa volcanic eruption as we pass the remnants of the destroyed island peaks, between the Sumatra and Java islands.

In 1883, it killed more than 36,000 people and injured 120,000. It unleashed a tsunami recorded to have gone around the world three and a half times. It was the loudest sound ever heard in modern history. People heard it 3,000 miles away in Alice Springs, Australia, and Mauritius Island outside Madagascar, Africa. It was 13,000 times more powerful than the nuclear bomb dropped on Hiroshima, Japan in WWII. Krakatoa released so many megatons of volcanic ash that there were amazing sunsets for the next five years around the world. In the USA, even Los Angeles, California, with its yearly average of 14 inches, had a record rainfall that year with an unbelievable 39 inches.

All of us are humbled and amazed by this Mother Earth we go spinning on every day, taking her for granted.

As we continue our drive to Jakarta, the driver asks me where I'm staying. I tell him I have no plans yet. He tells me we'll be arriving in the wee hours of the morning. Yikes, Singapore all over again.

Java

Jakarta

- Can't Find My Way Home -
Steve Winwood

Big hugs to the ladies. After we drop them off at the airport, Mr. Driver insists I'm his guest at his home. Yay! He and I head into the city suburbs. Pulling up to a nice-sized home, he wakes his wife. *Not the best way to meet the lady of the house, I'm sure.* She guides me to a small, but very nice, guest bedroom and my own bathroom with a western toilet **high kick** even a private entrance. I notice the western toilet bowl is installed the wrong-way-round, with the hole straight down and the shelf in the back... *curiouser and curiouser.*

They invite me to stay for the week while I'm here in the city. I tell them my friend is also arriving soon and they insist he'll stay too. I'm grateful to be staying with a friend in a city of 12 million people. Kuala Lumpur had 1.5 million, Singapore had 3.5 million.

When we pick up Jimmy at the airport, he's surprised and grateful we can stay with Mr. Driver. Down a few different freeways, we exit across from a Toyota billboard, and our new home is only a few blocks away. After he meets the family, we decide to get out of their hair and grab a taxi to find some dinner. Back in the hubbub of the downtown city, we decide on *Tony Roma's*? It's a barbecue ribs place from back home in California. How random to suddenly see something so familiar.

We enjoy a few days of goofing off around the city, checking out the sights, having some drinks, even renting a boat for a day trip out to 1000 islands, just north of Jakarta. It's filled with tourist resorts, so we lounge by the pool ordering cocktails all day—*it's Jimmy, remember?*

I tell him how cheap everything has been and how much money I still have left.

"Why not head over to Oz?" he says, meaning Australia. "You could find work no fuckin' problemo, as you Yanks say. You'd love it! The blokes'll love you!"

"Hmmm, maybe I should...?" I spontaneously plan to head to Australia. Why not? My ticket home is good for the entire year, the stewardess had said.

On Jimmy's last day he offers to mail my Asian pictures and post cards from Singapore as Indonesian mail usually gets lost. To kill some time just before his flight out, we buy some beers and walk through a huge flea market for souvenirs and people-watching. A man rides by on a bicycle with 25 live chickens all tied together, cascading over his front basket. They're not clucking or struggling, just staring at one another. A dog rides by on a little kids' shoulders on a moped. Food carts serve deep fried bugs and, yep, huge fish eyeballs. Sleeping grandparents that everyone just steps over. Metallica and Creedence CDs. I pick out a new Cranberries tape. There are faded, scratched, washed-out beer bottles filled with juice, with straws sticking out. Jimmy says they recycle here, *just not exactly the same way we do*.

I *love* this country.

Big hugs to Jimmy as he tries to get "Just one pash!" before he hops into a taxi. He said the Australian Embassy is nearby and I should get the required visa, just in case I can really go there. I ask a taxi driver at the flea market for directions. He agrees it's nearby and he'll take me. We ride for *ages*, and I'm so glad I grabbed a cab. It would have been hell to walk this far in the blistering heat. We finally pull up to the federal building. I'm exiting the car while he charges me an outrageous amount. As I'm handing him the money through the window, I see the same guy with the chickens, the washed-out bottles, the sleeping grandparents. The taxi driver snatches the money, speeding off.

I was dropped off at the exact same street corner where I'd been picked up. Ugh.

I *do not* love this country.

On my way in, I see a tourist guy sitting on the wall outside. In my arrogance, I speak English and ask if this is the right way into the Embassy. He looks startled, then smiles. This is how I meet Hujaf from Morocco. He's also getting a visa and we chit-chat in line. He speaks great English because he works nights in the tourist trade in Casablanca. I get my visa, no problem, but he has trouble and can't get one. He's really funny and we're both traveling east to Yogyarta, (*Joke-Jakarta*, everyone calls it *Joke-Ja*) We decide to meet in the morning and travel together.

I pick up some groceries for my hosts and grab another taxi—*oh, yippie*—to get back to my family's place. When the driver asks where to, I realize *I don't know where I'm staying*. Mr. Driver had driven me in the middle of the night and again on the freeways from the airport picking up Jimmy, to whom he gave the address. *Jimmy still has the address in his wallet.* We'd switched freeways three or four times in about a 20-minute ride to the edge of the city... I guess... North? East?

The taxi driver's getting impatient with me. *All I remember is the Toyota billboard.* And around the billboard there'd been some shops. *Ummmm, a motorcycle shop and gas station, maybe?* What if there's a Toyota billboard every three miles around the city? By now I've learned quite a bit of this language and I try to explain about the Toyota billboard, off on the right-hand side of the freeway. *Somewhere... in a city of 12 million people.* I can't think of any words. No words at all. I just keep lamely saying over and over "Toyota picture!"

I have my passport, but my backpack, money, and everything but the outfit on my back is in their home. I don't even know their last names. The driver starts driving, nothing looks familiar. He gets on the freeway, drives for 20 minutes, then heads back. I'm scanning every billboard, sweating bullets. The sun is setting. What am I going to do?

We're switching freeways again when an hour later, unbelievably, there's the Toyota billboard. I've never been so relieved in my life to see a car ad. I promise myself I'll always drive a Toyota. And it was of course, southwest. Close enough, right?

On the overcrowded bus with Hujaf, we really connect. People have brought their goats and chickens, some folks even sit on top of the bus. A small child sits on my lap until his stop. Hujaf holds the kids' mother's grocery bags while she sits on the floor with others. He's used to this kind of transportation, he says, and he tells me all about Morocco, his family, and some hilarious things.

When he was younger, he and his friends were pick-pockets. They had a special way of doing it, where they would talk to each other, right in front of the victims, about who they would rob on the trains or buses. They'd add the letters "N-C-R" to the middle of each word, kind of like pig-Latin in America. 'Pamencerela' would be my name Pamela. Then he would feel their pocket with his *knee* and keep both hands on his book to pretend to read so in case they felt him pressing against them, it would look like a mistake since both hands were visible. He would signal and his pal would rob them as they stepped off the bus.

I laugh at the pure shamelessness and ask him to tell me more things from home.

He's Muslim and once a year they celebrate Ramadan, their most sacred religious holiday. They fast each day for 30 days until the sun sets. He's a night tourist information guy, so he wakes up at 5:00 p.m., waits an hour until the sun goes down, then eats all night long, no regrets. He says it's all polite and respectful for the first few days, but after that, nothing runs smoothly. Everyone has fist fights and car accidents. They're *starving*. At the end, they celebrate by killing a goat and having a feast.

"Eeewww, you kill a goat?" I wrinkle my nose.

He looks confused, "Do you not kill a turkey for your Thanksgiving feast?"

"Oh, yes, you're absolutely right," sheepish as I realize, yet again, that everyone on this planet is the same wherever you go. *Technically, I don't kill it with my own hands, but whatever.*

Hujaf mentions men can't speak with a woman directly. If she's with another man, he must only speak to him. And women don't <u>ever</u> start a conversation with a strange man, hence his reaction when we met. I dig in with many,

many questions of my own. *Even if it's not your husband or father? What if you're out with your 11-year-old little brother and you run into your high school friends who've never met your brother?* Always the same. *How are you supposed to date? Find a wife?* It's still all arranged. I'm stoked to be from California, where I can talk to anyone I want.

We stop in a windy fishing village called Cilicap, on the edge of the thick jungle, we rent some bikes, tour a cave, then catch a boat to the next bus stop. We talk with other tourists, a shorter redheaded Australian man named Kaeden, a lovely Scottish woman, and a Japanese family. We're all exchanging stories and sights to go see, when I notice a very handsome Swedish man tape recording everything we're saying. He explains that he can really capture the feeling of the country by the sounds of the jungle, the energy people have from their voices, and he can make notes into it himself to remember how he felt. How cool! Video cameras are still expensive and bulky, while pictures *do* only capture half the story.

It's going to be a few more hours until the bus takes us to Yogya, so Hujaf, Sweden, and I grab some sodas and snacks from the local market. It's pretty dark out as we walk from the store, but we all see it in the same moment.

A bright red light flying 150 feet above us. It's not very big, but it's definitely hovering. It's weaving around in small, unpredictable circles. We all hesitate. Do we stay put? Do we hide? It's scary. We're all wondering the same thing.

Could we be seeing a real live flying saucer?

I spy a teenaged Indonesian boy riding his bike nearby us, so I stop him to ask, speaking in Indonesian, "What's that light?" pointing in its direction.

He watches it for a moment, his eyes getting larger with confusion and wonder. Then with a laugh, shaking his head slowly and pedaling off, over his shoulder he shouts in perfect English, "Radio tower!"

We all three roar with laughter, a bit embarrassed, grasping that it *is* quite a windy night.

Yet, it does not escape my attention that for just a split second, just a tiny pause, on this warm breezy night, four worldly people from Morocco, Sweden, America, and Indonesia, all looked up, held their breath and thought they'd spotted a UFO.

Yogyarta

- Wild World -
Cat Stevens

We arrive into Yogya about 11:00 p.m., say goodbye to Sweden, hot tamale Sweden, and grab a hotel. At this point Hujaf and I have become so friendly that I see him as a parallel-universe brother of mine. We decide to grab a bite, so I take the hotel business card and we walk.

The city seems safe enough, lots of tourists walking the streets at such a late hour. Coca-Cola and Marlboro billboards are my new markers, when Hujaf reaches over and punches me in the arm. Hard. *Hahaha, okay, funny guy*, but he keeps doing it. Six times. I laugh it off at first, then tell him to stop several times, but he won't stop. Eight times. *Now I'm pissed. DO NOT MESS WITH A PISSED OFF PAM.*

He does it again and I reach my left hand back, all the way from New York City, and I slap him across the face. It feels awkward, but I've got my camera in my right hand. He's instantly furious and I immediately understand his country's treatment of women means he can do whatever he wants, and he's never been slapped before. He holds his face, startled. Was punching me his way of *flirting*? Believe me, any and every man, ZERO women find this sexy. He grabs at my shirt with one raised fist, and, lucky me, a couple of young Indonesian men run over to help.

They calm things down, and Hujaf admits to all of us that, yes, he was trying to flirt with me. But when I slapped him, it was with my LEFT hand, and he's of those who eat with their right hand and wipe their ass with their left, so that's what made him so angry, that I touched his face with the wrong hand. *I'm American. That means I use toilet paper—maybe you've heard of it? This is a Grand Canyon cultural divide if you believe this is a date.*

I tell him I no longer wish to travel together, and again, lucky me, we'd only walked two blocks from our hotel. Both gentlemen walk us back and watch him as he grabs his bag and leaves the hotel. I'm so very grateful to these men as it hits me *(pardon the pun)* that Hujaf was a pick pocket. Maybe this was his way to get my bag?

My two heroes tell me of a recent fugitive serial rapist and murderer wanted by police in town, and they were glad to help. After many thanks, it's now midnight and I'm still starving. I head out for a bite, again. Like Las Vegas at night, there's a huge crowd walking the streets, lots of music, and flea market vendors hawking their goods. I grab some crackers and mango juice as a snack and wander the neighborhood.

Two silver-haired tourist ladies saunter into a courtyard filled with roses, along a massive, intricately carved fountain. If they can do it, I can do it. I feel safe enough. The ladies stop to read a post and I follow them, hoping to make new friends. I'm munching my snacks and reading the post over their shoulder. It's a *Wanted Poster* with a picture of that rapist/murderer.

Out of the corner of my eye, I see an Indonesian man staring at me from across the fountain. I shift my weight to show I'm uncomfortable with his stares. Finally, I turn to stare back. Still, he stares. I say loudly, annoyed, in the local language, "What are you looking at?" Like Julia Roberts in *'Pretty Woman'* when she's in the hotel room and the bell hop's just waiting for a tip.

He sprints around the fountain toward me.

I freak out. I run around the fountain as well, so he's still across from me. He switches and runs the other way, so I fall back and run the other way too. He keeps reversing, seemingly to block me from getting back to the main street. I hear a worried British voice, "Come with us!" It's the two old ladies, watching this unfold.

Oh. My. God. What is going on here? Is THIS GUY the rapist? Does he hang out by his posters, because I'm sure mostly women read them to heed the warning? And seriously, Queen Elizabeth, what are you going to do? You're 100 years old and unless you have a hidden James Bond pen/taser gun or have a black belt in karate, there ain't nothing you can do to stop this guy from hurting me.

I feel the sting of tears as no one even knows where I am. The last time I called home was maybe Jimmy's place on Batam a lifetime ago. I just checked into the hotel. No one would even know I was missing for at least a week, and they'd think Hujaf had hurt me after the scene we'd caused.

We've played cat and mouse around the fountain eight times when, out of the blue, I notice there are several off-shoots of alleyways leading out of the courtyard. I take a pure leap of faith, choosing the alley closest to me. I'm running like a nightmare of being chased by Michael Myers wearing that scary-as-hell white mask from the movie '*Halloween*.' I round two corners, then three. I look back, and he rounds the corner, too, still following.

I'm going to throw up, I am SO FUCKING SCARED. I'm not a fast runner and he's staying just far enough back to, what? Taunt me? Does he have a partner and I'm running into a trap? I can hear the crowded streets but I'm too far, plus I'm running in what has become an alleyway maze of lefts and rights.

I'm LOST. With a serial killer in pursuit.

The hair on the back of my neck is standing straight up. And he's still on my heels.

My heart's beating in my throat. I instantly get a flash of light guiding me to turn left, left again, and there, lo and behold, is my hotel. Clouds Part. Angels Sing. I look back and he rounds the corner just as I burst though the double doors.

I don't want him to know where I'm staying. The tears gush forth.

Through blurred vision, I see three huge 6'5" tourist men stand up rapidly. I've startled them. I haven't stopped running, even inside the hotel, and I run straight into the first man's arms. He has dark hair, big muscles, and glasses.

I've never seen him before. I don't know if he speaks English, if he's gay, or married. He's 6'5" so I don't care. He's clearly confused, but puts his arms around me tightly. I bury my face into his chest, sobbing, shaking with fear, while the other two are reaching to comfort me as well. I yell out through my sobs, "A scary man is chasing me!"

"You're safe now," he says in perfect English with a slight European accent. His friends get both night clerks and have one stay with me while the four men go out to check the scene. I'm sitting on the couch sipping some water, still trembling, when the men return moments later.

"Well, we found him..." says my savior. "When the four of us came out, he was leaning on the wall trying to look in. We all surrounded him, began yelling at him, pointing in his face. *He* started to cry. He was mentally... *slow*. The hotel clerk translated what we were saying, and he thought it was a fun game to chase you. We fibbed to him the police are coming. He was very sorry, saying he'll never come back."

WOW! My fear got the best of me from reading that damn poster. What a turn of events. My thoughts go to those English ladies who saw me running off down the alley and him hunting me. They'll forever wonder if I'm okay.

"I am Sandor, from Denmark," he says. The others introduce themselves too. Alex is also 6'5" and thin, with a mop of curly blond hair and glasses, reminding me of Tristan from Scotland in the treehouse in Africa. Liam is maybe 6'3" with a beard. I could not have run into—*literally*—a better bunch of men to save me from danger, perceived or not.

They've just arrived in town and are going to see the sights over the next few days if I'd like to join them. *Yes! Please!* And with that, we say our goodnights and I dream of Michael Myers landing a UFO, chasing me, and punching my shoulder. Hard.

Yogyakarta is a wonderful city. After that entrance I wasn't so sure, but over the next few days, these three guys and I have a great time sightseeing.

We arrive at Borobudur, the largest Buddhist temple in the world. It was built around 800 A.D., around the same time as some Hindu temples nearby, suggesting religions co-existing peacefully. Like the Egyptian pyramids, no one knows exactly who built it or when. It was abandoned as Islam took over Java and rediscovered under volcanic ash and thick jungle overgrowth in 1814. Still used today as a worship site, it was renovated and added to the World Heritage sites in 1991 and promoted to one of the Seven Wonders of the World.

It's *colossal.* Dark gray lava bricks in the step-pyramid model, 26,910 square feet. It's nine levels up with rows upon rows of over 1,500 intricately carved panels, and over 500 Buddha statues, yet nothing but an empty flagpole on the very top.

We've hired a guide to teach us. He explains the levels represent from lower life's desires, ordinary life, evolved life, and finally up to spiritually free lives. It represents the evolution journey you could take during one lifetime, but most will take over their reincarnated returns to life. Maybe we all start out like Charlie Manson and come back finally as Mother Theresa. He points out hints in the English sayings, "*I was in Seventh Heaven*" "*I was on Cloud Nine*" as well as a cat having nine lives.

Ancient carvings on the lowest level are of death, torture, murder. The next level up has "lighter" sins like greed, adultery, and stealing. Continuing up, even lighter sins like lies and gossip, until you get to that empty pole which simply represents Nirvana, the need for nothing but God. Not even the *desire* to know God. Pure peace.

He tells us it's a shame not to have been here at dawn, with no one here, just the mist over the jungle. Or a rainy day, when all the gargoyles have waterfalls pouring from their mouths. We're not to climb on the statues or point to them with our feet because that's offensive and downright illegal. We could go to jail for toe pointing. *Yep, pure peace.*

My Danish guys and I are running around each narrow tunnel, up steep, uneven stairs, enjoying each carving, taking lots of pictures and cracking jokes. *Are balding or getting fat sins? Spinach in your teeth? Bad hair day? What if you tell a girlfriend she does not look fat in those jeans? Sin, or evolving to Nirvana?* Seems to be a gray area.

Carved statues of the Buddha himself show 70 various meditation poses, the main ones facing the North, South, East, and West. Near the top, there are several upside-down bells called *Stupas*, with a Buddha statue inside. *Or they look like a being in an escape pod from the alien mothership.* We're told to touch his toe for good luck.

In 1985 Muslim extremists detonated 9 bombs here. *Who's mad at Buddhists?*

The view out over the jungle during the midday sun is awe-inspiring on this sweltering, muggy, Indonesian afternoon. We stay for hours and are still sad to say goodbye as we watch the sunset over the rainforest with our guide.

Next day, we head over to some of the Hindu temple ruins and the Sultan's Palace. The Hindu temples are fantastic as well, but nothing compares to Borobudur. We all kind of do the scene in the movie '*Vacation*' where Chevy Chase has just robbed the cash register and is trying to hide it from his wife, who is viewing the Grand Canyon for the first time. She says, "Isn't it just beautiful?" And Chevy bobs his head as if to say, "Yep, took it all in!" and rushes her off quickly. No disrespect to any religion, there are just so many temples all in the same area. We've seen plenty and hurry off to the Palace.

The Sultan's Palace is grand. Incense and tropical flower aromas drift through the open-air hallways, birds chirping happily, with soft Javanese music played by a small orchestra, creating quite a mystical, timeless feel. We tour it like a museum, viewing ancient weapons, art, vases, and historical pictures with the stories of the royal families who've lived here in the past.

Built in the 1750's, it has dirt floors throughout, and gilded ceilings pressed so lovingly by hand that I can still see each fingerprint. This represents the local spiritual belief that God communicates through your head, so as with Jesus, there are auras of light around each person. Above your head should be decorated with gold. You should never touch someone's head, including ruffling a child's hair, because it's taken as a great disrespect of their one-on-one communication with God.

Meanwhile, under your feet is on par with the devil and therefore, floors are made of dirt. As we recently learned, you would never point your toes at a person. It's considered insulting beyond belief. And while you're sitting on the ground, tucking your toes under your body is the right thing to do.

We stumble into a large room with a sheer paper wall in the center. On one side is a row of lanterns and wooden puppets with long handles, and on the other, 25 children sit on a mat, squealing with delight as a shadow puppet show begins. We join them for a while. What a great show, filled with lots of adventure and laughter as the puppet shadows tell a story of the Royals. Who needs modern movies when this entertainment is still providing excitement after all these centuries?

Through the gardens, Javanese dancers move elegantly, wearing feathers in their hair and costume makeup. The style of dancing is so unique to this side of the world, bending fingers backwards, showing their palms. Bells ring throughout the day as we stroll past large fountains, clear pools, koi ponds, and even an archery contest where the men shoot from a *sitting* position.

My boys and I are getting hungry as we pass a luncheon spread out on the floor for 30 monks. They're all dressed in bright orange robes with shaven heads and sitting cross-legged. They motion for us to join their feast. They're generous with their healthy vegetarian food, and they want to practice their English with us. I've learned quite a bit more of their language so I'm able to carry on a halfway intellectual conversation. Somehow the translator, I field questions about where we're from, which one is my husband, and when we'll have babies. We leave with full bellies.

That evening Sandor and his pals are leaving to Jakarta and it's time to say goodbye. He hands me a paper with a scribble of words with an '@' symbol in the middle, explaining that e-mail over his computer is the best way to reach him. *E-mail? What is he, some sort of scientist?* Never heard of it, and after he leaves, I innocently chuck it in the trash. I've given out my mother's home address to stay in touch, so we can connect the *normal* way.

That night out wandering by myself again, I run into the original two guys who saved me from God-knows-what with Hujaf. They mention there's a bus of tourists leaving in about 26 hours for an epic hike up Mt. Merapi, a volcano nearby. It's $10, so what the hell, I go with them to the tourist office to sign up right away. The bus leaves at 12:00 a.m. Yes, 12:00 A.M. After a 90-minute bus ride, I'll hike up four hours to see raging red lava that can only be viewed before the sunrise at 6 a.m., then hike about four or five hours back down for lunch before catching the bus back. I'm in!

It's about 11:30 p.m. now and Mr. Yogya and his pal take me for some tea and sweets to ask if I understand what an intense climb this'll be. "Sure, I'll be fine!" Hey, I took aerobics a couple of times a few months back before I began traveling, how hard can it be? He offers to let me borrow some camouflage pants and a sweatshirt from his Army days.

"Thank you!" I say, mentally thinking SWEATSHIRT? It's at least 105 degrees at night, but whatever. I don't want to be rude, and those pants sound perfect. With all this lovely rice-rice-baby, let's just say my only pair of *loose* jeans have become my only pair of *tight* jeans.

That next day is filled with writing letters home, walking around the shops, and doing some laundry with a bar of soap in the sink. I must relax before getting up at midnight. Mr. Yogya drops everything by and wishes me luck, and we agree to meet up the following day, so I can return his things.

I've been staying up past midnight daily, so I get the genius idea that *it will be easier if I just stay awake all night long*. Because that always works out.

There's a knock at my door from the bus driver, and I'm out the door with my side-attachment from my backpack wrapped around my waist like a fanny pack. It's filled with a few crackers, a bottle of water, my passport and some money, my camera, with a last-minute choice to grab some lip balm and my trusty pen-flashlight. I'm wearing my friend's pants, a T-shirt, with my thin long-sleeve shirt tied around my waist, socks, and boots. I decide he's crazy about needing the thick sweatshirt, so I leave it in my room.

I'm *ready*.

On the bus I meet 15 other tourists from Australia, Hong Kong, England, Italy, and Japan. We all chit-chat for a few moments and then try to get some sleep until we arrive. There's a small café at the base camp, and we snack on some tea and jaffles before we begin. We chat a bit more, excited to be on this adventure. Everyone seems to have been training for a few weeks, and one guy's even a triathlete.

Two guides who climb with us—*our Sherpas?* —clap their hands to get this party started. We all head out the door to the path behind the fence and out into the darkness. In the light from the dim bulb in the café, I can barely see anything beyond the fence... just thick jungle, huge boulders, and a tiny pathway. The first guide is leading us all and the second guide is last.

Okay, actually he's behind me, because I'm now last. *Wait, what, how am I last?* Less than 10 minutes in, the meandering path has suddenly become quite steep and I'm out of breath. The others are fading into the distance with their guide leading them. The other guide, Maday, becomes my personal chaperone. Every step I take is like lifting my full leg up as though I'm climbing a steep stairwell. *How long is this hike? Where did everyone else go?* Half of those people are in their 50's and 60's. *When do we get a break? What have I signed up for, exactly? Dios mio!*

I settle into a snail's pace. Up the hillside, we can't even see the flashlights of our team or hear them talking anymore. Maday has a flashlight, so I see how useless my penlight is, with its tiny blueish tinge, and I shove it back into my pocket. We agree to hold hands since I can't see one thing in the darkness. Soon the jungle starts to thin out, the boulders becoming plentiful. The air's less humid and the path underfoot becomes looser, more squirrely, with moving rocks. It's downright slippery.

I ask for another break. We stop to sit on some rocks, while I sip some water and lightheartedly complain. He smokes a cigarette and I notice he's wearing flip-flops. The cheap flip-flops that are, like, 75 cents from Thrifty.

"How often do you do this hike, Maday?" I ask, in his language mixed with mine.

"Every day," he replies between long inhales.

"You do this hike EVERY DAY?" I cannot believe it.

"Yes, every day, I have a family..." he says.

"Wow! I'm a lazy girl!" I make him laugh.

"Okay, time to go," he chuckles.

I grasp his hand again and step directly where he steps, one foot after the other. I try to be cocky and look up at the stars. Pitch black night with every star from the entire Milky Way splattered across the sky like a Jackson Pollock painting. My ankle rolls as I slip on a rock, so I instantly regret my choice. He steadies me, I'd better focus. Step, slip, steady, step, slip, steady, again and again. My thoughts wander like I'm in the Stephen King story '*The Long Walk*,' which, even if you've never read it, the title kind of gives away. It becomes a meditation moment. I think of how insignificant I feel again, like an ant climbing up his ant hill home. I think of the big cities I've lived in and visited and what the people I've met must be doing now, at this very moment. I miss everyone and no one.

My lungs are burning, but I prefer this feeling to the strenuous burning in my thighs and knees. Hell. This is becoming hell. I should have '*Buns of Steel*' once I reach the top. I'm requesting breaks in rapid succession, every 15 minutes now. Maday's patient with me, and I feel the need to let him know this is quite the challenge for me as I'd broken my knee last year in a snowboarding accident.

This is not accurate at all. In reality, five years earlier, when I was 19, both of my knees popped out of their sockets as I was learning to snowboard. Yep, an old friend Mark had taken me to the top of the mountain on my first run ever and put me on a regular board. I learned two things in that moment. That I needed a "goofy-foot" board, not a regular board, and that I didn't know how to *STOP*. I was so pissed off that day because I'd paid big bucks for lift tickets for me and my then-boyfriend. I decided to keep on going the rest of the day and ended up with soccer-ball knees for the next couple of weeks. Totally worth it, until this moment. But I feel the need to explain why I'm so extremely slow.

I'm requesting breaks every few minutes now, and Maday's prodding me along, reminding me that in order to see the hot lava correctly, we must be up on the rim before sunrise. I have blisters forming on my toes, and I'm sure I have a rock inside my sock that's gently turning me into the devil.

"Okay, this halfway point. We go now," Maday politely puts out his cigarette. I scowl. I was hoping to distract him into a longer conversation so we could really rest. It's been two and a half hours straight up and yay, we're halfway. *Is it too late to cancel?*

Hours ago, I'd put on my long-sleeved top because it was a lot chillier with the wind whipping alongside the mountain. Holding hands and stepping into Maday's footprints, I'm still laughing that he did this hike yesterday and will do it again tomorrow. In flip-flops. Smoking. I'll bet he could compete with Olympic athletes.

Maday and I haven't spoken for ages because the air's much thinner now, colder, and I'm in serious pain. Everywhere. I think my eyebrows even hurt. This is torture, and I paid $10 to torture *myself*. I'm an idiot. I would pay $1000 to get a cable car just to go back down the mountain at this point, *lava-shmava*. This is why I've never seen lava in my life before and will not be seeing it anytime in the future, that I guarantee. *Step. Slip. Step. Breathe*

I was a mouth breather from the first 10 minutes, so hours in, my tongue is swollen. I can hardly swallow, with dusty layers in my mouth, in my eyes, and all over my face from each step I take. Inside my nostrils are frozen and the ice-cold air is making my entire face numb. My tiny, thin, long-sleeved shirt is a joke and back in my room, I have the gift of a warm, thick, cozy sweatshirt from a thoughtful local who knows this mountain like the back of his hand. I'm a *dummy*.

At last it's proper break time. Maday has a picnic lunch for us and we sit in a flat area for about 15 minutes. The rim is still about 35 minutes up. I eat and drink slowly, my mouth works as well as can be expected, like *after a root canal*. Maday, thank God, insists we sit for the full break, explaining the blood in my legs and brain need the rest. *Whatever you say, dude!* But now that we're not moving, I'm so bitter arctic frozen to my bone marrow that I cuddle up to him for body heat and he cuddles me right back. This poor man, up here, this cold every night. My heart goes out to his family and I hope they know what a hard worker he is.

I feel hands shaking me. I must've fallen asleep.

"Okay," Maday lifts me to my feet.

I'm disoriented for a moment. I dreamt about my Dad. He and my mom divorced a few years back and I hadn't spoken to him in a year before this trip. I was angry at some of his choices and would always pass the phone over if I heard him say hello. *Rude and immature? Yep.* But the last time he called he caught me and said how proud he was as I was leaving to Africa the next morning. He said he loved me, and I said it back. In my dream he was talking to me, but I couldn't understand him.

Maday smiles, "Lava time!"

Great. Lava.

This had better be diamond-studded lava shooting out of a unicorn's ass. More monotonous marching, with a hint of a glow beginning on the horizon. We pass groups from my bus coming back down.

"It's totally worth it!" Italy shouts.

"Just beautiful!" says Japan.

I'm telling my fib about my broken knee snowboarding to everyone we pass. I guess I didn't learn my lesson from the fib to the French people about my math genius. *Why am I so pathetic?* So what if 90-year-olds are passing me? Or if 3-year-olds are cartwheeling up ahead of me?

I make it to the top.

Maday's still holding me steady as I peer inside this monstrous volcano's rocky mouth. The scorching steam blows in my face, immediately soaking my hair. It smells of rotten eggs, like a breakfast belch from a trucker on Route 66. Sulfur, I'm told. Hundreds of feet below, there's bubbling mud, spewing, shifting, red hot lava. It's neon orange. It seems to be a gushing torrent, yet the stream keeps changing its path, as its walls cannot contain it. The lava just repositions wherever it wants.

I. LOVE. IT.

And I'm warm.

I'm filled with wonderment. I feel like a little child peering at the burning embers of a campfire. This is magical. I know in my raw, aching bones that not many people would have done this hike, and not many will in the future. This grueling march was worth it. This planet is stunning and mysterious and I'm lucky to catch even a glimpse of it.

It's difficult to look away, but as the horizon glows a bit more, I turn to take in the view.

We're above the cloud line. Reminds me of landing in Kuala Lumpur. A colossal blanket of ice-blue sapphire cotton candy clouds, with just a handful of mountain peaks poking through, as far as the eye can see in every direction. The wind is strong up here, and I'm extra glad for Maday's presence. He reads my mind and says thoughtfully, "It as beautiful as first time I see it, even though I see every day."

We smile at each other. I'm elated to be without my group. He'll never remember me—*well, maybe as the slowest girl in history*—but I'll never forget him. We've shared a journey through the time/space continuum. We're in another dimension. A realm above his city, above the island of Java, floating in the sky above the country of Indonesia, just the two of us, on the edge of the inferno.

I bend down and pick up a few Javanese volcanic rocks, slip them into my pocket as a souvenir. *Yes, I'm stealing. I'm a liar and a thief tonight, for some reason.*

As we head down the mountain, the sun's beginning to rise and I can finally see. Now it really looks like I'm on another planet with the loose gray moon rocks. In any direction, I see miles of white clouds and other mountain tops off in the distances. It's still ultra-windy, magnifying the chill factor. As we walk through the clouds, it looks like the end of '*Raiders of the Lost Ark*' where the spirits are flying all around them after they've opened the Ark. The clouds are moving rapidly by as they're sucked up and over the mountain top we've just come from. No jungle in sight yet.

After an hour, we run into an English family that appears to be older parents and a teenaged son, also heading down.

"Oh great!" I shout to them, happy I'm not last anymore. "We'll walk down with you!"

They're pleased to see us, and we all chat the rest of the way down. Marion, Timothy, and their son Frederick are traveling though Java on their way home to Nottingham. It looks like the mom is struggling with her knees, so lucky me, we slow down our pace to match hers. Going down sounds easier, but because of the slippery loose rocks, I have to regain my footing with each step or twist my ankle.

We agree this is the most adventurous thing we've all experienced. Four hours down the mountain, stopping for water breaks and not much else. The sun is up high in the sky and my long-sleeved shirt is around my waist as the heat warms my frosty bones.

We trek through the jungle and finally enter the café. The family says goodbye before heading off around the side of the restaurant, jumping into a car, and speeding off. I'm so confused. Where are they going? My entire bus is at the tables ending their feast. Everyone starts clapping for me now that I've arrived.

"We thought you'd fallen into the volcano!" Italy shouts. Everyone laughs and they tell each other the fib about my broken knee again.

"Sorry we were so long," I say recounting how we'd just met a family and walked them down because I thought they were on our bus.

"No, mate, we've been waiting hours for ya... our entire bus was together this whole time!" says Australia. "Seeing as how Mt. Merapi just erupted and killed 50 people coupla weeks ago, we were about to send out a search party for ya!"

Oooops. The British family wasn't even with my tour. *Wait, what? Mt. Merapi just erupted?* On the bus ride home, I learn Mt. Merapi's elevation was 9,737 feet. No wonder it was brutal. And the translation literally means *Mountain of Fire.* It's been active regularly since 1548 and, since 1987, it kills people every few years. Lovely. Maybe I should have—oh, I don't know—*ASKED ABOUT IT BEFORE I GAVE MY $10.* I'm a dummy... told you.

Finally, back at my resort, I sleep forever and wake up the *following* day. A note is slid under my door from the front desk clerk regarding returning my friend's pants whenever I wake up, no worries.

*****

I call home that evening and talk to my mom, telling her everything about my cargo boat ride, the elephant jungle trek, Borobudur, and now the volcano.

"Have you talked with Dad lately? I dreamt of him while I was asleep on top of the volcano," I say.

"Um, actually there is something going on with him..." Mom says, awkwardly. "A couple of nights ago, your Dad was driving himself home after dark and his tire blew out on the freeway. I guess he was tired or something because he got out to change it, took the old one off but did not put on a new one, just jumped back into the car and drove off..." she pauses to let it sink in.

"I don't... get it... so, what does that mean, exactly?" I say slowly.

"The metal rod where the tire should've been was scraping along the ground causing sparks to shoot out as he was driving along the freeway at normal speeds. The police chased him for a while before they realized there was a problem, and pulled in front of him to slow him down to arrest him. They're detaining him now for 72 hours on a psychiatric evaluation hold to see if he's a danger to himself or to society," she says with great care.

"Oh my God! Is he all right?" I'm overwhelmed at this news. I'm standing in a phone booth facing a crowd. I feel claustrophobic and turn away, wishing I had a chair.

"He's fine. We've spoken while he's been in jail. He said he was just very tired," she continues, exhaustion in her voice. "My dearest Pam, I don't want this to affect your travels. He's fine, we're all figuring this out, but I don't want to alarm you, okay? You continue on your journey and I'm sure this will all be worked out when you do return in a month or so. I don't want you to worry..."

"Oh... okay... I guess..." I glance out to the people waiting to use the international phones and one arrogant asshole lifts up his hands in a rude way, pointing to his watch and raises his eyebrows as if to say 'I own this place and it's my turn!' Probably American.

There are five other phone booths. I waited myself and I've been on for maybe eight minutes. *Kiss my ass.* I frown, turning away. "Are you sure, Mom? You know if you needed me to come home, I would hop the very next flight back to LA..."

"I'm sure my darling, now tell me where you're going next!" She's changed her tone and wants it to end on a good note, so I tell her Bali is right next door, and I've looked into the possibility of Australia as long as I still have enough money. She's excited for me and we send our love and kisses.

A few days pass, just healing my legs, window shopping, meeting travelers as they pass through Yogya. When I return the pants and sweatshirt, I mention Bali and my pal's response was, yep, "Bali is more dangerous than Java, be careful..." I just wink now. All over Indonesia, this is their *thing*.

Bali

Kuta Beach

- Alabama Song (Whiskey Bar) -
The Doors

Just as with my other bus rides, the seat on this bus folds flat—all the way back—so I can lie down, which sounds awesome... until the guy in front of me folds his down into my lap. I've no choice but to put mine down, too, even though I want to look out at the lush rice paddy landscapes.

A red-headed girl jumps into the seat across the aisle and introduces herself as Jane from Canada. She's also traveling alone through Java, and we bond immediately. She tells me she just finished getting her skydiving license and I'm fascinated. We talk the entire bus ride, including the ferry ride over to Bali. There are now clusters of other travelers as Bali's quite the popular destination. We share a mini-bus ride over to the beach town of Kuta. We're both weary after 15 hours of traveling time, so I ask if she wants to share a hotel room.

"Can't," she says with a smile, "I'm meeting up with an old flame tonight. Cory from England."

"Oh, hubba, hubba!" I grin, "No worries, it was lovely to get to know you."

"Why don't you meet up with us tomorrow on the beach and we can grab a bite?" she volunteers.

"Sounds great!" I say as she guides me to the popular area of backpacker's places on Poppies Lane Two. I grab a nice room with a double bed, a ceiling fan AND a table fan, my own private bathroom with a proper toilet, and my usual tea and banana pancake breakfast—all for $6 a night. *Yes, super fancy.*

Next day, I'm walking down to the beach to meet Jane, passing outdoor cafés one after another. They're filled with tourists and show Hollywood movies while they eat. Everything from '*Superman*' to '*Beverly Hills Cop*', '*Goodfellas*' and '*Forrest Gump*.' Locals everywhere hock their wares on the street. Silver jewelry, tie-dyed shirts, snacks, and whatever else you might need or want. It's an entirely different vibe from being off the beaten track for so long. I think I need this for a couple of days.

"Braid you hair?" an old Indonesian lady asks as she pets my hair.

"Um, no thanks," I reply, smiling, remembering J.J.'s advice.

I notice everyone along the sidewalk smiling as I pass by, they are the friendliest people, I must say.

"Water?" asks a man selling ice cold bottled water.

"Yes, how much?" I'm happy not to have to look for a store on the way to the beach.

"$9," he says, smiling.

I already know they're 10 cents at the market, but he needs to make a profit as well and I've been bargaining since I got to Asia. There's an Indonesian 'local price' which I'll never get, being a traveler. It's fine, I have more money than they do. There's the 'Caucasian price' and the outrageous 'Japanese price.' This man just made up his very own 'Oprah Winfrey' price. So here we go.

Big smile. "40 cents?"

"$8," he says with a nod and a smile.

"45 cents?" I pull out money, smiling. *They're small bottles.*

"$5," he says smiling, not taking the money.

"Okay," I say nodding, "50 cents!"

"$4!" He nods and smiles.

I hold steady, "50 cents."

"$3...?"

"50 cents..."

He smiles and shakes his head no.

I smile, put my hands together in a slight bow and shake my head no as well and walk away.

"Okay, okay, special price for you!" He calls me back smiling. "$2."

I smile and walk away again.

"Okay, okay, 50 cents," he's still smiling so I think we both know he's been caught trying to gouge the crap out of me. Who falls for that? I hand him the money, he hands me the bottle.

"Actually, I'll take two bottles," I prepare more money. Going to need a hotel bottle too.

"Okay," he says getting more water. "$9."

Seriously? We literally just did this. "50 cents." Smile, and we go the full round about again including me walking away twice before he says okay to the 50 cents again. God bless him for trying. I hand him an extra 50 cents to say, "Good game!" and off I go with my bottles of way overpriced but amusingly memorable water.

Kuta Beach is awesome. Surfers in the water. Palm trees lining the beach. Everywhere I look, people are getting massages in little cabanas, manicures/pedicures, drinking piña coladas from raw coconuts. Lots of beer and shots. Party beach. Everyone on a moped. People selling necklaces, sarongs with suns and fish all over them, so flippin' cute. Girls back home would *love* these. A kid with a machete in his belt loop climbs up a palm tree to get me a pineapple for 10 cents. He's skinning it with his machete so quickly I think he should be missing a finger or two. He hands it to me upside down so I can eat it like an ice cream cone.

And yes, the beaches are topless. *Told you.* I notice immediately because there are two chubby surfer girls out in the waves floppin' and bouncin' away. I'm not so brave. I might consider lying around but floppin' or bouncin'— *never.*

I'm already really tan from my travels thus far, so I decide to keep my girls covered. I would have pasty white triangles, then sunburned, neon red triangles shortly after. Maybe later.

As I spread out my sarong, I hear Jane call my name and I wave her over, along with the blond buzz-cut guy Cory. We introduce ourselves and laugh, cracking jokes and travel talking.

We're just lounging in the sun and climbing in and out of the sea, when we notice a 6'6" Howard Stern look-alike walking toward us. He has long, black, unruly hair, tied back in a low ponytail topped with an '*Indiana Jones*' leather hat. He's walking up to us with a handful of books he's looking to sell or trade. I notice the New York accent.

"Hey, you're American? Me too!" I say. We invite him to join us and we all introduce ourselves. He's Sean from Manhattan and he has a couple of the funniest, most amazing travel stories I've heard yet.

He'd been in the Melbourne Osaka Cup yacht race, on a boat that smashed up onto the rocks and he was rescued just as it sank. He'd been held prisoner in a Nepalese jail for three days, falsely accused of marking his passport. After refusing to sign their documents and shouting only "American Embassy!" he was released for a trial. The officials thought he'd try to escape to the border, but Sean walked straight to the American Embassy. The guard had the sweetest Texas accent he'd ever heard in his life. "*Nothing can touch you here, son, you're on American soil now!*"

"Hey," Sean finishes his stories, "You've just got to meet my new friends Lucy, Weston, and Perry—they're farther down the beach. They're British and have been here in Indo for a few months now. I'm meeting them for dinner. You guys want to come out with us?"

Cory and Jane bow out because they must fly home the next day, but I agree to meet up. As the evening sky turns purple and orange, we agree on a place and head back to our hotels to get ready. I can't help it—I buy a couple of new gauzy sundresses for a night out. One is ivory with vertical stripes and the other a deep maroon with pink tie-dyed stripes for $4 each. It feels grand to have something new after six months with the same outfits.

I'm starting to get a little lonely for a boyfriend. I didn't feel a spark with Sean, but maybe this Weston or Perry? I'm open to falling for someone. Wouldn't that be a romantic story? We fell in love halfway around the world, in some exotic location? I choose the ivory dress and put on some mascara for the first time in months.

Lucy is lovely. I know we'll be fast friends. She has long brown hair and is quick to smile and laugh. We could pass for sisters, even though she's a Skinny Minnie. She's been suffering from typhoid for months and is finally feeling better. She was on another island of Indonesia and when she fell ill. They moved her into a convent with chickens and goats roaming the same halls with people cooking stews. Nuns kept trying to stick her with *dirty needles.* Thank God for Perry and Weston, who never left her side and kept insisting on new sterile needles. When I ask why she didn't fly to Singapore, I learn typhoid's a food-based illness, and you can't fly to any other city that might be better equipped to handle the disease because *your guts will explode at high altitude.*

Perry's charming, sort of formal with curly red hair and a sweater vest with a bow tie, mannerisms like C-3PO from Star Wars. No spark there. Weston has long hair and a cool vibe but no spark with him either. He's Lucy's brother and they have a great story from their childhood back in England.

Lucy lights up when she tells it, "Our mother was the drama teacher at the local theater and always had extravagant costumes up in the attic. Late one night, as bored teenagers, Wes and I decided to dress up as Angels and head out to the old War Memorial statue at the curve in the road by our home. We draped ourselves in flowy white cloths and huge feathery wings. We attached golden halos and brought baby powder and glitter to squeeze up into the air, so it all floated slowly back down over our heads. We attached flashlights to our belts, so it looked like we were lit from within as we struck Angel poses as cars drove by."

Weston continues, "Well, it worked. Too well, and the few cars that drove by swerved to get a better look. Suddenly, a huge bus drove by, and one woman couldn't believe what she saw. She ran from her seat to the back of the bus, plastered her face and hands to the back window to get a better look at the Angels. Powder! Glitter! Flashlights!" They start giggling at the memory.

Lucy finishes, "After the bus drove off, we got bored and headed home, laughing. Three years later, our Mum was at a party and met a woman who said she was forever changed by her 'spiritual encounter.' Turned out her entire life had changed when, three years earlier, she'd seen Angels at the old War Memorial. Our Mum didn't have the heart to tell her the truth!"

Yes, we'll be fast friends for sure.

We all head out to dinner and dancing at the only nightclub in town, an open patio bamboo place called The Sari Club. In the wee hours of the morning as we exit the bar, gangs of children rush us chaotically. They hold up newspapers, as if to sell them, but in reality, their small hands are rifling through our clothes to pick-pocket us.

There's no shame as I grab a kid's arm out of my purse and shake my things free from his hand. He just shrugs and walks back to where he was sitting with his other friends, waiting for the next drunk people to come out. He's about 10 years old.

A loud motorbike roars by and we all turn to see one moped carrying five children, all about seven years old. One kid sitting on the handlebars, three across the seat, with one standing on pegs on the back tire, holding his buddies' shoulders. No one's wearing a helmet of course, all are barefoot, each kid's carrying a machete, and yep, going down a one-way street the wrong way. *Ummmm, where's your mom? Are you allowed to have machetes? Where are you going?*

Indo at its finest.

Lucy, Weston, and Perry are heading north in a few days to the town of Ubud, to visit the Monkey Forest, and Sean and I are invited to join them. Yes, please! Over the next few days still enjoying Kuta, I meet with my new crew and have a blast swapping travel stories. I tell them all about Africa and cool things I've seen around Asia. Lucy's traveled quite a bit, and she even lived in a kibbutz in Israel for a while. She would like to become an underwater scuba documentary film maker. My eyes fill with heart shapes. *Coolest dream job ever.*

"In all my travels it's so rare to meet an American," Lucy tells me. "Especially one traveling alone and even more rare to meet a girl on her own. Before Sean, the last American I met was months ago, when I first arrived in Bali. He was a California surfer with a long blond ponytail and a little trimmed beard, and he hated TV..."

I immediately burst out laughing, spilling my drink from my mouth. "Was his name Jay? J.J.? That's the guy I was with in Africa!"

Everyone laughs. "No way!" says Sean. "There seems to be a small group of this world that travels, and you'll keep meeting up in different countries. It's happened to me, too, I love stories like that!"

Fun days meeting up with them for lunches, beach days, and shopping. Dinner in cool outdoor cafés and dancing each night at The Sari Club. Each night, even my little burglar buddy waves to me, smiling, of course. We're pals now. In my hotel room, I get a knock on the door... locals asking if I need a massage or if I'd like a jungle painting with the tribal lives of a time long past, which took him two months to paint, such detailed work. Yes!

The Balinese can't believe I can speak some Indonesian, so it surprises them when I can whip out some slang or a joke. They love it. They're so used to the Australian tourists there that the Australian accent is the only thing I'm hearing. "G'day mate!" is now the local Balinese greeting. Probably as unexpected as it is for foreigners to come to California and see some of the Mexican nationals saying, "Hey dude!"

It's clear a lot of ex-pats and foreigners have come to Bali to live, build a home, a career. Lots of people might have come over for a vacation, fallen in love with the sights, maybe a Balinese native, and built a family here. Businessmen sipping coffee, as well as the board-shorts tourists sipping beer.

Nothing like living where other people choose to vacation.

We all travel south to Nusa Dua, with its surf breaks and resorts. We climb the steps at Uluwatu, the Hindu temple carved into the cliffside rocks overlooking the beach. There's a band of small children playing local drums and plinky-plunky magical music.

I can't wait to see more of the island when we leave tomorrow for Ubud.

Ubud

- The Monkees -
The Monkees

Vibrant verdant rice paddy fields tiered down the mountainside allow us to fall in love with Bali. The conical woven hats worn by the Balinese keep the sun off their faces. The long sarongs, flowy shirts, and flip-flops are what typical locals wear. Ox-driven carts are filled high with freshly chopped banana bunches, mangoes, and pineapples. We pass thatched huts with no walls–just four sticks and some hammocks—along with modern homes and vacation villas. Small bird houses for spiritual offerings consisting of fresh flowers, fruit, and burning incense out front of each home line the dirt streets. Nude families bathe at the town well on the side of the road. Toddlers play in the street using Coke bottles as pull toys.

In Ubud, we pass groups of locals making massive papier-mâché creatures, like Trojan Horses. They're for the next funeral cremations in a few days. One red dragon has a gargantuan penis, half the size of the entire beast, with a long streamer hanging out of the tip. *Yes, we must make sure it's a fully phallic funeral.*

Since the others have already booked a room at a fancy hotel, Sean and I decide to share a room around the corner at an open-courtyard, tropical-garden kind of place.

"$10 per night, plus tax," the happy clerk tells us, pointing to the printed pricing sign above his head.

"Okay, we'll take it for $6, and we won't be paying the tax," Sean replies, smiling. The clerk smiles as they settle on $7, nodding his head in agreement.

Wow, I've learned a bit more, not having bargained with my hotels, thinking they were more professional. I'd already noticed—even if I explain I'm in a rush, or I explain I already know what the price is—that I can't get the best price without bargaining. A few times in a rush, I've just paid the first quoted price and they're actually insulted, they seem to *like* the bargaining. For me it was fun at first, but now it's getting tedious.

Everyone smiles at us as we walk around town. Stray dogs, cats, people on mopeds, and bikes are everywhere. Women practice their dance moves on their patios, teaching the young girls how to bend their fingers backwards, in what they consider to be the most beautiful thing a woman can do.

We all hike to a place for some birdwatching, which turns out to be nothing more than a tree in front of a house where a guy comes out, asks us for a dollar each, then he goes back into his house. Zero birds land the entire time we're there. We laugh at the rip-off as we walk back through jungle trees with vines hanging low in the canyon, passing random statues on the side of the road, maybe ancient, maybe new, there are no descriptions.

That night, half asleep, I get up to use the bathroom facilities. I must be dreaming. On the wall, over the western john, is the largest gecko I've ever seen. I literally rub my eyes with my fists to make sure it's still there. *Is that the Tick-Tock Croc from 'Peter Pan'?* Its head is the size of my open hand and its body is longer than my forearm. I usually think geckos are cute and I'm not scared of reptiles, but I decide to wake Sean. After all, Komodo is only three islands away, famous for their Komodo dragon lizards that grow as large as a car and eat goats and children.

"Pssssst, Sean... you have to wake up and see this!" I shake him.

"Hmmmmm? Yeah? What's up?" he asks, still half asleep.

I hear a *Crunch-Crunch* sound coming from the loo. "Quickly, the world's largest gecko! You've got to see the size of this guy!"

Sean walks over with me and I gently nudge him into the room.

"What the hell is that?" He yells, fully awake now.

Crunch-Crunch-Crunch! There's that sound again. I peek over Sean's shoulder—*Okay, okay, he's 6'6"*—I peek under his arm to the wall where Godzilla was a moment ago. It has moved above the mirror, and where it had been walking, the plaster has fallen off the wall with every step. That was the crunching noise I'd heard. This guy's huge feet grabbing *chunks of wall* and shattering it to the ground. Laughing, we scare him away for good.

Monkey Forest is stunning. It's a small national park with meandering stone pathways through the dense leafy jungle. Hanging vines, small stone walls holding back ferns, carved statues of Hindu deities and animals. The path leads us by miniature mossy ponds, ending at an exquisitely carved Hindu Temple. Every few steps I have to navigate through monkey families. Mommas, daddies, babies, big ones, little ones, fat ones, skinny ones, funny ones. The largest seem to be the size of a golden retriever. The funny ones are picking at each other's hair, checking out their nuts, hitting each other, stealing human sunglasses, drinking out of a soda can and, sadly, even smoking cigarettes.

As I roam the gardens and watch the monkeys play in the sun, splash in the water, cuddle their families, and wrestle each other, it really hits me that I'm so far from home. This couldn't be farther from Los Angeles and I love it.

One baby monkey comes up to me as I'm sitting on the wall. He twists his head, blinking his wide eyes. I offer some banana, and he jumps onto my lap, grabbing my hand to share in holding the fruit. He's like a baby drinking from a bottle of milk. Out of nowhere, a large Daddy primate elbows him off my lap and jumps up himself, so now I'm feeding him the banana. A little scary, but he's insisting. He looks into my eyes and I glance down a bit, smiling but not showing teeth. I remember my friend Scott's adventure overseas.

Scott's a military friend stationed in the Philippines, and one day, at the barracks, he was heading back from a shower. Still wrapped in a towel, he spotted a giant long tailed macaque monkey at the jungle's edge, staring directly at him. They shared a moment of eye contact, so Scott smiled at him, to show he was a friend, that he was peaceful. The monkey went *BAT-SHIT bonkers*, beating his chest, screeching at him. Scott ran back to the showers, towel flying in the wind, the monkey in hot pursuit.

Friends sitting at nearby tables could see Scott sprinting through the camp, clutching his towel, with an enormous angry monkey chasing after him, furiously screeching. The roar of laughter filled the moment. Scott ended up in the last shower stall, hiding in fear, while the enraged primate peeked behind each shower curtain, still looking for him. It finally gave up and left Scott alone, not realizing he was in the next stall. I guess showing your teeth to a monkey means you're in attack mode. *Ooops.*

*****

Exploring Ubud further, we participate in a cremation burial party. They burn the Trojan-dragon-penis-horse we'd seen, complete with dead body inside, carrying it around town, singing and dancing.

Now that's my kind of funeral.

I hear of another type of funeral practiced in the more tribal areas like Suluwesi, an island to the north. It's called a 'Sky Burial' where they lay their deceased out on a rock table and let the birds peck at the carcass until it's nothing but bones. It sounds gruesome at first, but the more I think of it, how wonderfully spiritual to feed the animals and be closer to Heaven in the birds' flights each day.

Hindus believe in reincarnation, so while they are sad and will miss their loved one on this plane, they feel strongly they will be reunited with them in another lifetime.

My dear friends head to Denpasar to fly out.

Alone once more, I catch a ferry to the next island.

Lombok

Gili Trawangan

- Enjoy the Silence -
Depeche Mode

There's a second Kuta Beach to the south for surfers, and that's where I find myself after climbing off another heavy metal minibus. There's no one around, except bats flying overhead while I eat my lonely dinner. After only one night, dreaming of my sister and mother back home, I decide to go north to the tiny group of islands called the Gili's. I'd heard rumors of these islands, with no phones, no lights, no motor cars. Sounds perfect.

A full five hours later, I arrive at the dock and hop on a canoe filled with drunken tourists. Lucky for me they get off at the first island, Gili Air. A few more people get off at Gili Meno. I've had enough of being ON the beaten path and am holding out for the third and more isolated experience again.

I arrive at Gili Trawangan and fall in love. It's a tiny island I can walk around in two hours, with crystal clear teal water and white sandy beaches lined with palm trees. Transportation on the island is limited to bicycles or horse-drawn carriages. Naked local children playing in the dirt with nothing but rocks, but grins so large they melt my heart.

I walk up to a row of beach huts for weary travelers. I pay for a week in full and prepare to relax, snorkel, and get to know the locals. I open the door to my new place and am so pleased to find light wood furniture with white sheets, a mosquito net, ceiling and table fan, a bathroom all to myself and of course, my own front porch with a hammock on the waves. That includes the banana pancakes and sweet tea I've come to see as my staple breakfast, all for $3 a night. I shower up and realize I'm bathing in salt water. After drying off, I'm surprised I still feel as clean as a whistle.

Ukulele strumming floats from beyond the next porch over as I head out to my hammock to watch the sun set and have a listen. I'm swinging away as an Indonesian man with copper skin, almond shaped eyes, and long black hair down to his bum, walks over plucking his mini guitar.

"You Australian?" he smiles, nodding as though he knows the answer is yes.

I shake my head no, smiling, making him guess a bit more.

"England?" he asks, sure of himself this time, still playing away, taking a seat on my porch steps.

I smile, enjoying my new game. Shake my head no again.

"I know, I know, French?" he smiles.

I give in and tell him the truth, "American."

"Oh? I don't meet many Americans here, but I have many Australian and English girlfriends. What is your name?" he asks.

"Pamela," I reply. Maybe he thinks I'm traveling alone to find an Indonesian boyfriend. "And what is yours?"

He stops playing the ukulele, gazing deeply into my eyes. "Midnight. Bob Midnight." *Bond James Bond.*

I try to hold back a laugh at this smooth talking—and very obvious— ladies' man. "Oh, an Indonesian name?" I'm making fun now. We shake hands then he goes straight back to his song, which might be '*Hotel California*.'

"Yes," he says, suddenly very serious. "I take you to dinner now."

"Oh, okay, thank you." I'm happy for a new pal, one who's a local, but not sure when I should drop the bomb that I have a *pacar*, a boyfriend. I'll buy my own dinner.

We walk down the beach, both of us barefoot, strolling slowly toward the Christmas lights dangling around the only two cafés on the island. They're right next door to each other and offer exactly the same menu. Instead of a competitive feeling, there's a friendly vibe, with all the tables spilling out onto the sand, candles aglow. Stray dogs and cats roam through the chairs looking for scraps. The sky's turning tangerine while 70's music plays softly.

Bob Midnight pulls out my chair on the sand, still warm under my toes, and orders for the both of us. I'm not sure what I'll be having but I'm up for a food adventure again, having forgotten my earlier food flops. Fresh lemon-lime fish arrives with bottles of beer and water. Yum. The conversation flows as he asks me all about California and tells me what it's like growing up on such a small but touristy island. He keeps hinting that an American girlfriend would make his friends jealous. People pass us, waving hello.

The next table over fills up with eight travelers, just getting to know each other. Bob already knows them all and introduces me to an English blonde beauty, Naomi, and her boyfriend, a Danish blond beauty as well, named Hans. They have multiple new pals... Japanese, Argentinians, and Kiwis. After the generators shut off loudly, leaving only candlelight on the entire island, we all stay up until the wee hours, laughing under the miles of Milky Way starlight. Bob walks me home, and with a slight bow, he heads back up the path in the direction from which he'd come.

I sleep like a baby to the sounds of the small waves lapping gently against the shoreline, just steps away. The humidity is perfect with just a slight sheet covering me. I dream again of my mother and my sister calling out to me. I wake up thinking of them, but I brush it off. After all, I just called two weeks ago, and I've been calling home once a month.

For the next few days, I enjoy my new pals. We've all agreed to meet up for snorkel days on the beach, breakfasts, lunches, and dinners. On the beach, Naomi and I get a good chat, "In Denmark, Hans just didn't understand my not wanting to be considered a slut if I slept with him on the first date. They just sleep with you if they like you there. It hurt his feelings when I moved his hands away a bit. Not like in England!" She laughs as she admits she slept with him on the first date anyway but felt guilty about it. He just couldn't understand her guilt. She feels a freedom to live how she wants now that they've been together for three years.

Bob plays his ukulele for us each day. We snorkel, swim, eat fresh sea food. Hans had nearly been stung this morning when he picked up his shower scooper and found a mama scorpion with a nest of babies.

And then just as swiftly, everyone's hopping onto the boat to exit back to the other larger islands. Bob also has some things to attend to out of town. I still have a few days left and just want to sit and stare and think and read and sleep with the kind of silence rarely enjoyed in life.

I'm truly relishing in the travel lifestyle. There are always new pals to meet, new places to see, new foods to try, new adventures to enjoy. I don't know why anyone ever goes back. I trade in some books for new ones at the local bookshelf where I can take one for free if I leave one. I buy a small hand-carved turtle trinket.

I decide to walk around the entire island and see what I can see. After 20 minutes of walking along the beach, I seem to be crunching along on top of rows of broken coral. A reminder that the locals here fish with *dynamite* and blow up everything to smithereens to get their daily fish intake, not realizing coral takes hundreds of years to grow. Now I'm walking on top of what looks and feels like miles of piles of bones.

After a few days, I thought I'd feel like Robinson Crusoe, but I just feel lonely and blue today. All my friends are gone. Maybe it's time to head back to Bali, then on to Australia. I still have seven days left before I'm thrown in jail for overstaying my visa, but I haven't yet arranged a flight anywhere. If I keep going it will be Australia, because I already have my visa there from my time in Jakarta. Or else I can buy a ticket back up to Kuala Lumpur and hop on the next flight back home to Los Angeles. I smile. I already know which way I'm going. G'day mate!

That night I dream *again* of my family. My mother and sister were on the other side of a glass window trying to get my attention, but I was busy, not paying attention. This is weird.

I investigate making an international call, only to be told there are no such facilities on this tiny island. I'll have to wait until I get back to Lombok.

Hmmm, I'll give it another day or two and if no new friends show up, I'll head back.

Senggigi

- Disappointment -
The Cranberries

"Do you have any friends with you?" My mother hasn't even said hello, she's given a deep, troubled, sigh. I'm on Lombok, in the west coast city of Sengiggi, with easy travel back to Bali.

"No, I just left some friends. It's Dad, isn't it?" I feel it in the pit of my stomach. The dream of him while I slept on top of the volcano, him driving off without his tire, and my recurring family dreams to call home. And now, it's in my mother's agonized tone.

"Yes, Pam. It's... Dad..." her voice breaks. "I am so very, very, sorry to have to tell you like this. He passed away one month ago!"

I blink in slow motion.

The wind is knocked out of me. I can't take a breath. My eyes sting but the tears don't roll out. My heart shatters into intricate, complex emotions.

This man raised me from three years old. He was my stepdad, but I grew up calling him Dad. I remember their wedding, up on top of a hill in San Francisco, where grasshoppers kept jumping up under our long skirts. Bagpipes played and someone gave us girls a sip of beer. He helped me with my homework, he put Band-Aids on my scrapes, he taught me how to work on cars, he took me camping and played with me. He built us a white club house in the backyard. We filled it with two of those little roller beds that mechanics use under cars, and with hard hats and teacups. It was called Sandy and Pam's White House.

"What happened?" my voice is just a squeak. *He pretended to pull candy from my ears.*

"Well," Mom says, "After you and I talked three weeks ago now—and I had to tell you the story of how he spent a few nights in jail because of the tire—I called in to check on him and they'd released him early. He had a dental appointment, a root canal. He went home after and it looks like he'd just collapsed. They think it was a heart attack. Pam, he was there for a few days before anyone found him. I feel so deeply saddened by this. I wish he didn't live alone during that moment..."

"Are you okay?" I ask. "How are the others?" *He loved Jim Croce and Dire Straits.*

"A part of me is now gone. You know, even after a divorce, you always think you'll see that person again, graduations, weddings, grandbabies... I thought we'd still share our lives, just in a different way..." Mom takes a deep, exhausted breath. "We've already buried him. Your friends went for the service and I had to inform everyone that you didn't yet know of his passing. He had a military funeral because he'd served in the US Navy, a Vietnam War veteran. He was sent off with the full honor of a 21-gun salute. You'd have been so very proud."

He was 49 years old.

Colossal salty tears roll down both cheeks.

"I'll be home on the next flight," I state calmly. I'm nauseated. The air feels too thick, too humid. Time feels slow, everything surreal. If I catch the earliest ferry in the morning, and minibuses in Bali, a 23-hour flight from Denpasar, I'm sure I could get a flight direct instead of going up to Malaysia? Depending on availability, I might be back in LA in maybe 36 hours?

"Now, wait... we've all been talking about this. I am so sorry, but... you've actually..." she hesitates, struggling for the right gentle words, "...*missed* the grieving family time. You missed the funeral. You missed all the bonding and reminiscing. We're all back to work, to school, and the rest of our lives. If *you* need to come home to be with us for *your own sake* then, of course, come back, but really think about this. You'll come home and sit here and watch TV as we all go about our daily lives," she pauses. "Oh, Pam, he actually passed away the night you dreamt of him while on your volcano hike. It seems *you* were the most connected to him spiritually, even so very far away, my dear."

I talk quietly with the rest of the family and they offer their condolences as do I to them. My mother gets back on the line and we agree to talk tomorrow, after things had sunk in a little. I'm sure I'll have more questions about it all. I need to lie down at my hotel. My mouth is dry. I'm so sleepy.

I missed everything.

I never thought about that. It was a given that if my family needed me back home, I wouldn't hesitate. I would fly back immediately. We never thought about what to do if I were needed, yet somehow missed everything. I'd originally planned three months away, and now this journey is turning into something else entirely.

"Hey there! I know you!" I hear a Canadian voice. I turn and there's a man I've never seen before. He sees the puzzled look on my face and tries again. "You were on my ferry? On the way over a couple of weeks ago. How have you been?" He's a tall blond, smiling, full of energy, loving this travel life.

I. Can. Not. Do. This. Right. Now.

"I'm so sorry. I can't..." I start bawling. Bawling in the street in front of a happy stranger as motor bikes and buses whiz by. As little kids on bicycles ride by with fruit in their front buckets. As women weave baskets and hats from the palm fronds at the local market across from the international phone call bungalow. As stray dogs and cats and chickens run around the local street vendors cooking their fish-eyes and serving up papaya juice in plastic baggies with a straw from God only knows where.

"Hey, are you okay?" he softens, genuinely concerned.

"I j-just f-f-f-found out my D-d-dad d-died!" I stutter between sobs.

He reaches out for a hug. "I'm truly sorry. Just now? You just got off the phone with your family?"

I can barely manage a nod.

"Where are you staying? I'll take you back." He's kind enough to walk me to my "hotel." It's really a hut with a mattress on the floor, a mosquito net, a shared bathroom down the path. Zero bells and whistles.

This kind friend brings me some fresh water. He asks me what happened. And I tell him everything. I mean *everything*. Like in the movie *'The Goonies'* when the kid Chunk started all the way back from kindergarten when the bad guys demand he spill *everything*.

I talk for hours. He offers to go get some dinner for us and comes back a few minutes later with fish, rice, water, and even some chocolates. I talk about my Dad *all night*. I literally talk and talk and talk. I tell him about the White House, about the arguments while I was a teenager and my parent's divorce six years earlier. I tell him how we were in a fight for a year and that last time we spoke I'd said, "I love you" before my trip. At least I'll always have that. I tell him about my dilemma now, do I go home or stay?

As dawn breaks, my friend must leave to meet up with friends and hike a volcano if I'd like to join. *Hell No!* But he'd love to meet up again in Bali in a few days' time to check on me. As the door closes behind him, I understand he is a Human Angel. I cannot believe the kindness of such a well-placed stranger in my life. Thank you, Angel, sent directly from my Dad.

Sorry for talking your ear off. I don't even know your name.

Hours later, I haven't eaten my beloved pancakes. I cannot believe my Dad is gone. *My Dad is Gone.* I call Mom again and we talk for $420.80 worth of a collect long-distance charge. She pays, of course. She suggests I take the rest of the week until my time is up in Bali to decide for myself to keep going or to come home. It's not an emergency to rush back. My family encourages me to keep going, telling me how proud my Dad was of my free spirit and my adventurous nature. Mom seals the deal with the thought that *maybe Dad made sure I'd missed out just so I would keep going.*

Funny thought. Here I am wracked with guilt that I'd been snorkeling while my family was suffering. That I'd been drinking beer while they were crying. I should've been there. I should've known something was wrong. I'm reminded that I *did* call home a full week earlier than normal, because I felt something was off.

I have a lot of thinking to do.

After the call, I sleep all day and around midnight I'm starving. I walk out into the strange city of Sengiggi to find some food. It's not touristy at all in this neighborhood. The only thing I can find is a little old woman with *those* black teeth. I've since learned it comes from constantly nibbling on Betel Nut. She's cooking something on a street stand, under a single flickering streetlight bulb, dangling dangerously from a loose wire. It'll do. I don't feel like bargaining and just pay full price of a whopping 75 cents for what might be a fried Spam and grape jelly sandwich. Doesn't really matter. I don't even care. I find one wobbly chair and rusty metal table.

As I pick up my sandwich, I'm feeling so lost. The loneliest I've ever felt. Not just on this trip around the world on my own, but the *loneliest in my life, ever*. I don't want to be alone. I just wish I had a friend, *any* friend, to eat with me at this very moment.

Out of the corner of my eye, I see a stray cat dragging a very large and very dead rat in its mouth. It's heading out of the alley and straight toward me. It runs directly under my table and stops. I'm not sure what to do. I glance to the cook lady for help but she's lying on the ground, trying to get some sleep. I look at the cat. We lock eyes. Just as I'm about to prod him along with my flip-flop, he rips open the rat's stomach, spilling intestines onto the ground in a pool of blood. The cat looks up at me, I swear he's smiling, as if to say "Quite a score, huh? Let's eat!"

A moment of pure joy washes through me on this desolate night that *this cat* is the friend I'd just been wishing for. Dinner buddies. So, I take my first bite of this "greasy pork sandwich served in a dirty ashtray" as per Chet, in the movie '*Weird Science*.' The cat and I eat every last bite, even licking our fingers and paws together. I feel we have a real moment of connection, as he keeps glancing back over his shoulder, possibly to ask me "How can you eat *that*?"

The overwhelming smell of the fresh rat blood is my cue to head back to my "room." Stay or Go? Go or Stay? These thoughts are with me, I can't sleep. I get up, walk across the street to sit on the beach, watching the moon rising higher into the sky. It must be about 2:00 a.m. and I'm sobbing on the warm sand. I'm listening to my tapes. I choose that new one, the Cranberries, and one of the songs is about her Dad, she's asking him '*Why?*' My Daddy's Dead. *He called me Pammy Peanut and took us camping.* I feel a wave of nausea, but it's deeper than just the Spam burger fried on a dirty pan in the middle of a third world country or witnessing the rat guts spill onto the ground.

I feel a hand on my back. I jump a bit, and a small Indonesian man sits next to me asking if he might practice his English with me. I know I should be nice, but I don't feel like it. I say firmly in Indonesian, "Please do not disturb me, I have a boyfriend." He bows and leaves me in peace. I feel bad.

I sleep until noon. I walk to the large resort on the edge of town, the one on the beach with the vanishing edge pool and water slide from the mouth of a mammoth stone Tiki statue. Easter-Island-like for sure. I'm going to have another day like the one in Kuala Lumpur, when I stole that swim on the rooftop of that fancy hotel. I grab a fresh, fluffy resort towel from the resort bin, spreading it over the pristine resort chair, between the resort hammocks and the resort pool table.

I'm going to stay all day. I laugh to myself that I might put another meal on another room number. Watch it's even the same couple, still traveling, and I do it to them again. I unwrap my dress and jump into the deliciously cool, refreshing black-bottom pool. I'm under water when I see a shadow at the pool's edge, someone from the hotel, wanting to know what I'd like to order, perhaps. I hear him trying to get my attention. I ignore him as I come up for air and immediately go under again.

He's still hovering, this time I can't ignore him, he's kneeling down in my face as I come up for air.

"You guest here?" he's glowering.

"Pardon?" I pretend to have a British accent, for some reason.

"You guest here?" he speaks loud and clear.

"Yes," I say with a smile.

"What you room?" he's not buying what I'm selling.

"272," I answer confidently. Was that the same room number as before? I'm thinking I might just order something to make it more believable.

"We no have rooms like this!" steam comes out of his ears.

"Oh..." I'm caught red handed.

"You go NOW!" he barks at me.

"Okay," I say, defeated, climbing out. Lucky for me, he has to walk all the way around the pool, which gives me enough time to grab my stuff and run.

You think that was funny, Dad? I know you did! That's my new job, making you laugh in Heaven as I clumsily float along this big blue marble?

My amusement makes me consider continuing on to Australia. I still have over $1,500, so I might as well head south for another month or two. Then, with the last of my cash, I'll fly back up to Malaysia to catch my remaining flight home. I'll be gone 12 months instead of three. What's the difference? My Dad's still gone. I missed everything. It dawns on me I can talk to him whenever I want, from wherever I am. I instantly feel closer to him in this moment of stealing a swim than I have in years.

"Hey Dad, you okay? Thanks for the cat buddy last night. I'm feeling like you're with me on this journey now. Want to go home or to Australia?" I'm muttering to myself as I walk back to my mattress.

I turn the corner into my No-Star Hut and see three smiling ladies.

"Hi, I'm Annie," says Annie, a beautiful brunette from Scotland with hazel eyes and a wide Julia Roberts smile.

"Hi, I'm Pam," I shake their hands, explaining I'm from Los Angeles.

"Hi, I'm Eva and this is Carrie, we're from Australia," says one of the two tall and tanned blondies.

They've just checked into my hotel and were thinking of getting a fancy drink at a resort if I'd like to come. Yes, please! I highly *don't* recommend the Easter Island one as I explain what just occurred down to the south. We all laugh at me and agree to head to the north of town.

Soon we're all settled in with our strawberry daiquiris. It comes up that I'm thinking about heading to Australia, but just learned about my Dad's passing. Everyone hugs me and offers their condolences. The two Aussies share that their fathers had died as well. One just three months ago and the other a year earlier. They tell me it gets easier to handle the idea of it as time passes. We all end up sharing lovely memories of growing up with our families. *I needed this. Thanks, Pop.*

They're all heading up to the Gili Islands tomorrow, so I share my adventures there and how cute it is, to look for Bob Midnight, and how he's harmless. And to give candy to the local little girl outside the market, the naked one who just plays with dirt, but sings and giggles all day long.

The Australians keep telling me about where to go. *I was just asking my Dad about continuing on to Oz, when around the next corner, there are two ladies encouraging me to go to Oz... Coincidence?*

Maybe this is what it's like to have a new Guardian Angel watching over me in my life? I'll stay alert to subtle signs.

We stay up half the night having laughs and sharing hopes and dreams. As I've noticed, the people I meet traveling delve into deeper conversations much quicker than regular friendships back home. Maybe because we know we have limited time to get to know each other. Let's do it right the first time.

I listen to the Cranberries again, the song called '*Disappointment*' with the lyric "Then" over and over, but with her Irish accent it sounds like "Dad... Dad... Da-a-aaad..." The whole album really hits home. The tears come again. Good thing I've got that little teddy bear my mother had given me for a cuddle. I'll call him Bob. Bob Sunset.

Bali *...dua*

Kuta Beach *...dua*

- Ode to My Family -
The Cranberries

I catch the next ferry to Bali. The girls were lovely, and I needed that. Now it's time to really focus because I have only three days before I go to foreign-third-world-jail if I don't figure out a flight anywhere.

The ferry's crowded and I fall asleep in my seat, emotionally exhausted. I'm so out of whack from all these midnight meals, then sleeping in...

The loudspeaker jolts me awake, screaming Indonesian instructions of some sort. Everyone's up on their feet, chaotically grabbing at life jackets, tying them on. *There is not another one for me.* Oh. My. God. *Are we sinking? Have we been hit? Pirates?* No—seriously—pirates are everywhere in these waters. The voices are still earsplitting. *Titanic? Iceberg? What is going on here, people?*

After about 10 minutes of this, the voices are much calmer, everyone takes off the jackets and sits back down. *What the hell just happened?* Still no idea, but next time, I'm grabbing a life jacket first.

My old room is available on Poppies Lane. I head out to walk around and I'm in for a surprise. All the bars and restaurants seem overly crowded with drunks and I don't feel like talking to anyone.

I pass some native women selling sarongs with fish and suns on them. They're so cheap, I'll buy some later to send home. The girls in California have never seen anything like these. They'll go crazy for them, like those tie-dyed skirts selling out so quickly back at *The Gunston 500*. Maybe I could start a business, bringing things back for my friends?

Should stay or go... I grab something weird and wonky to eat from a street vendor, again not feeling like meeting drunk vacationing tourists.

As I sit on the curb snacking, I see a man from the ferry commotion this morning.

"Hey!" I shout and wave.

"Hey!" he smiles and comes over.

He's tall, dark, muscular, and handsome, with a twinkle in his eye. His name's Tafale. He says he's a Samoan, but grew up in New Zealand.

"So, what was all of that commotion on the ferry?" I ask, "Was it serious?"

"Oh, that was just a training drill," he laughs.

"Oh, so no big deal then? I didn't get a jacket and couldn't understand what was going on. Scared me to death!" I giggle.

"Indonesian boats sink all the time, so they just started practicing what to do. Probably safer in the long run..." he smiles. *All the time?*

We end up talking for a few minutes and head to the beach, sitting in the sand under the half-moon. After a bit of conversation, he tells me his Dad had died about five years earlier. I tear up and tell him I just found out mine was gone a few days ago. He offers warm condolences about how, as I begin the acceptance stage of grieving, I'll be able to focus on just the good memories. We hug and part ways as I go back to my room to read some more. I can't help noticing yet another person put in my path, who also lost his parent, is encouraging me to continue this journey as long as my family is supportive.

I'm in the travel agent office at the beach to see about a flight to either Australia or Malaysia. I'll leave it to fate. The answer back is nothing to Malaysia for a week and same with Australia, except a flight that leaves in three hours to Darwin. A week. That puts me over the limit for sure. And three hours—Holy crap, Batman! I need to clear my head for a moment. I go sit in the sand again.

"Hello again, Pam," that happy, warm voice is Tafale, as he walks up with a 10-cent pineapple, sliced neatly by the kids with the machete. I tell him I'm not sure what to do next. He sits in the sand with me as I go over everything. It's been hard, not being able to bounce ideas off a friend. He insists I take the flight to Australia, and says he's headed there himself soon.

A woman approaches to sell me some sarongs, and I realize there's no time for that now. I'll have to give up on this dream of importing and exporting if I buy that ticket. I barely have time to pack.

"No way. Run to grab that ticket and come right back to buy the sarongs. There ***is*** time," Tafale takes charge of my flailing indecision.

I do as I'm told and go next door, grab the ticket to Darwin and come back to Tafale, who has kept the Indonesian saleswoman waiting for me. She begins to show me her selection, one by one.

"How many pieces total?" I ask, not in the mood for *Bali time*.

"95 pieces," she smiles.

"I'll take them all," now I smile. She blinks as though she doesn't understand.

Her friend comes over to explain, and she also has 80 pieces herself. I buy those too. I feel like a high roller for sure. I send Tafale next door to the tourist shop to get me a duffel bag to stuff them in.

I take 175 pieces in all. It takes only about five minutes of quick decision making and *Voila!* A new career when I get home to LA? We shall see, but for now, I'm so grateful for the extra nudge from my newest friend. I zoom to my hotel, pack up quickly, and check out.

At the airport, I'm surprised to see Tafale in line at the counter for my flight. He changed his flight, and I'm so excited to have a new travel buddy. We rearrange our seats to sit together. He's trying to get to Cairns but, like me, he just has a ticket to Darwin. I feel a million pounds lighter as this choice to continue on was handed to me from each person I'd met. After hearing the circumstances, not one person said it would be better to go home.

Australia

Darwin

- Wherever I May Roam -
Metallica

Dad? Can you help the flight land safely? Thanks...

I enjoy getting to know Tafale on the flight. He grew up in New Zealand as a Kiwi, splitting his time with Western Samoa, sharing the heritage and traditions of his people. He's been living on and off in Australia and wants to travel a bit before heading back to New Zealand.

We land in Darwin and since it's a short two-and-a-half-hour flight, we step into the night air that's as hot and humid as Bali had been. The energy of the city is much different from the island lifestyle I'd become used to. It feels alive and exciting, but in reality, it would compare to Lancaster, California. About 100,000 people, with nothing but desert for days in any direction.

The Northern Territory is known for Kakadu National Park with its wetlands and aboriginal rock paintings dating back as far as 20,000 years. The aboriginals might just be the oldest human tribes in the world, living in these regions as far back as 50,000 years ago. They believe a giant Cassowary bird egg cracked open and out spilled the universe. The Rainbow Serpent then sang into existence rocks and mountains, oceans and rivers, trees and wildlife.

Tafale tells me to stay on the lookout for Miracle Berries. They're known just in the region and if you eat a Miracle Berry, while it tastes nice, the real power is that it changes your taste buds so the very next thing you eat tastes extra sweet, extra juicy, extra delicious. My mind will be blown, I'm told.

Tafale finds us a hostel that sends a shuttle to the airport, and we grab a spaghetti dinner there for $5. *Heavenly* after not having Italian food for months. What's even more heavenly? Not having to bargain for the price of the dinner. It's just $5. They say "$5" so I hand them $5. Easy. End of transaction. None of the song and dance routine that's required for all purchases in Asia. I don't have to do the steps of pretending to be a student, walking away, getting called back. Fun, yes. Interesting, of course. But exhausting after months of bargaining daily, hourly, no matter what the purchase. And, yes, my first dinner in six months that does not contain any trace of white rice.

We talk about buying bus tickets to Cairns when Tafale notices a job board looking for workers for the next few days at the airport. Interesting. We agree to check it out in the morning. We're stuck in bunk beds in a room filled with drunk, smelly men. I forgot about not having my own cheap Asian bungalow. I feel homesick, but for Singkep, Bali, the Gili's.

After a breakfast of Muesli *(Oh, like granola, yum!)* and Vegemite toast (*Oh, bitter brown ear wax, nasty!)* the guy behind the counter tells us I'll have a better chance of getting work if I get a Tax ID number, to pay taxes properly into the country's system. So, I open an account at some Australian bank and get the ID number. At the bank, I get a "debit" card, whatever that is. Something I'd never even heard of. I can use it at restaurants, the grocery store, even taxis. It's so easy, I just type in a "PIN" number. I can't believe we don't have this in the USA. It needs to be a top priority for sure.

Back at the airport, the job is to wash airplanes. Little two seaters, larger mosquito flights that carry 20 people, and even your big boys. The boss is a short, deeply wrinkled 75-year-old man who's pregnant with triplets apparently, just *all belly*. He looks me up and down, "Think ya can handle this, little girl?"

Ummmm, I used to be a roofer, jackass. "I'll try," I smirk.

Tafale and I are to sand, wash, and dry them, get them ready to be painted. He leads us to the sandpaper piles, the sponges, and buckets. Yes, I'm the only girl in the midst of 15 guys, but so what? I've certainly washed a car before. *Who knew airplanes even got dirty?* I begin sanding a small one. It's physical, but I get into the groove of it. The men begin to relax around me and start cracking their man-jokes. I laugh. I've got a few man-jokes of my own. Tafale has his own plane and we race to see who'll do a better job.

Hours later, the boss grunts, "Tucker break!"

My muscles notice I'm no longer relaxing in a hammock. They're ahead of my mind and throw the sponges down. Burgers... yuck. I order a chicken sandwich at least. Fast food tastes *weird* after having rice, fresh fish, fruit, and veggies for months. It tastes *chemical.*

The boss is checking me out as my shorts and tank top are soaking wet from the soapy water. I look like washing-planes-turned-wet T-shirt-contest. As we finish up, Boss Hogg shouts out, "Back to work! You, Pam, clean up everyone's lunch!"

Oh, I thought you were just a jackass, but you are a Sexist Jackass. My mistake. Why doesn't everyone clean up their own lunch, hillbilly? I clean up at a snail's pace, realizing I'm still earning the same amount as sanding the underbelly of the plane. And this is easy. I literally walk a cup of soda to the trashcan, then go back for a napkin to throw away. Then another cup, just one thing at a time. *Talk to me like that, I'll show you.*

We endure for several days and earn enough to buy a plane ticket instead of a bus to Cairns, in the north east of Australia. As we climb aboard a 747 jet, I settle into my window seat and can't help noticing the wing of the plane, just covered in dust and soot. *That is one filthy, filthy wing.*

Cairns

- A National Acrobat -
Black Sabbath

I'm surprised. The landscape reminds me of Southern California. Cairns is more of an industrial, touristy city. It's along the northern bit of the Great Barrier Reef, but a trip out to see the natural Wonder of the World is a three to five-day trip for big bucks, not just a few hours for reasonable bucks. Tafale and I are at a hostel again, sharing bunk-bed rooms with young travelers. These places are called *youth* hostels and won't allow people over the age of 26. I'd always felt that I wasn't quite old enough to live my "real" life yet, and now that I'm living it, I'd almost missed it altogether.

There are a few jobs offered up on the hostel board. 'Picking Bananas' sounds interesting, until a girl who works there tells me about the *daily tarantulas falling onto her shoulders*. "You just gently knock them off your shoulder or your head. You get used to it, I swear!"

Ummmm, no thank you. And I may never eat another banana in my entire lifetime, thank you for telling me that story.

The other ad is for 'Circus Workers.' Hmmmm. Like, a real traveling circus. Would I be a ticket agent, or would I be a trapeze performer in the tiny, shiny outfit? Knowing my luck, I'd be cleaning up the elephant dung and my only friend would be the bearded lady.

Tafale's a scuba Dive Master trying to get a boat job. I'll try for restaurants. We separate during the days to search. His sounds more fun, but we're both having no luck.

One afternoon, Tafale and I are swimming at the hostel pool, when he asks what religion I am. I tell him my heritage and background but basically say I'm a happy hippie. I ask his in return.

"Taoism," he replies.

"Is that like Buddhism?" I really don't know.

"No. It's like Taoism..." he says, also genuinely.

"So, you're Muslim?" I don't understand.

"No, I'm a Taoist," he's amused now. He thinks I'm messing with him, but I've never heard of Taoism.

"I'm sorry, I don't know what you're saying. What is Taoism?" I really want to learn.

Tafale rolls his eyes, smiles, and dunks his head under the water.

"I'm sorry!" I laugh, "It's not my fault. We don't learn that much in the US about the rest of the world, just our own history, and maybe a bit of Eur-o-pe-an!" I'm hollering as he swims away.

He swims back and tells me that I *have* heard of it. He quickly explains the belief system, of keeping mind, body, and spirit in harmony. In balance.

"Do you know the Yin/Yang symbol? The circle with the wavy line down the middle, half black, the other half white, the dot in the center of each?" He points to a Billabong surf shirt hanging over a nearby chair, one with the symbol on it right in front of my eyes.

"Oh, yes, of course! I do know it!" I'm not a complete doofus. Mostly.

"You know there are about 4000 religions, right?" he asks.

"Of course! Actually, no. I guess I just assumed the big ones took over. Christianity, Judaism, Buddhism, Islam, and, I don't know, Hindu...?" I think of '*The Simpsons*,' with Hindu Apu shouting at Homer "*There are 700 million of us!*" I know I sound doofus-y again. He laughs.

"Well, I'm sure *you* picked the right one!" I poke fun.

As we dry off, he tells me he's been invited north to Port Douglas and he's leaving in the morning. I find myself a little jealous and I wonder if he's visiting a girl, maybe an ex-girlfriend. I'm bummed that I'm not invited.

I feel truly sad to say goodbye to Tafale, but give a heartfelt hug as he hops onto his bus. He changed my life on the beach just a few days ago, inspiring me not once, but twice. He talked me through my father's death, helping me decide to keep traveling, and also, to purchase those sarongs to maybe become a real-life importer/exporter someday. I'd already mailed the duffel bag full of clothes back home, telling my mom that I would be continuing on my journey. Maybe I had some help from above.

I instantly get a flash of Byron from New Zealand, whom I met back in Jeffrey's Bay, South Africa. He'd mentioned if I ever made it to New Zealand to look him up, but also that he had friends to stay with in Cairns if I ever made it out here. I'd forgotten all about the phone numbers until this moment, because I hadn't considered my trip would go beyond Africa—let alone a world away.

I decide to call the number of Byron's friend, Brock. He seems laid back on the phone, and he's happy to hear I'm a traveling buddy of Byron's. He says I could stay with him and his roommate, and he can pick me up in an hour. Yes, please!

As I'm waiting at the front desk, I see some men's steel-toed black work boots on the floor and ask where I could buy some like those because mine are disintegrating fast. The desk clerk tells me these were left behind a month ago and he was just about to throw them out if I want them. I try them on and—Boom! Free boots for me. My size exactly. *Gross, yes, but a mini-miracle, yet again.*

"Are you Miss Pam?" I hear a deep voice behind me as I'm sitting on the floor trying on my new kicks.

I turn to meet Brock, a tall, fit, shaven-head, goatee-having guy. He grabs my pack like a gentleman, and we head to his place, talking all the way. Turns out he and Byron had been Surf Camp buddies for years. His apartment has an extra room with a queen bed, and I can stay as long as I want. Brock's roommate, excuse me, now that I'm in Australia, his *flatmate,* is Tanner, and he's friendly as well. I guess they travel a lot and can appreciate having friends in strange cities. And Brock is going camping this weekend if I want to come along.

My first night there, Brock and Tanner take me out to dinner, play pool, and meet their group of friends at a cool beach bar that shows surfing videos on TV. It's fun to meet the group and everyone asks me what it's been like—a girl traveling on my own—what America's like, about my family and friends back home. I tell them this feels a lot like home to me. I surprise all, especially myself, with my turn at the pool table. I break, then proceed to hit a ball into the pockets six times in a row.

Brock and Tanner have to work all day, so I decide to explore their home. I'm standing in the kitchen scoping out their bookshelf and grab the first thing that catches my eye: The Joy of Sex. I'm flipping through, fully enjoying the hilarious flabby 70's bodies, the creepy hairy guys and, of course, who doesn't want to try the hot tub handstand?

"Hey there, Miss Pam," Brock has snuck up on me. I hadn't heard the kitchen door open.

"Ummmm... uhhhhh..." I'm embarrassed for the first time in my life, caught being a pervert. "Just reading an educational novel," I grin, showing my pick.

"Oh, you perv!" he laughs.

"Hey, it's on *your* shelf! You're the perv!" I laugh back.

"It's Tanner's," he says, blushing as well. "That's my story and I'm sticking to it!"

Laughing, I close the book and we pack for the camping trip tonight. Since I couldn't really find work, I'd discussed with them where I should head next and everyone agreed on Airlie Beach, eight hours south.

Brock and I drive south a few hours, pull deep into the forest, and take out sleeping bags. No campsite. No tent. Just going to sleep on the ground, next to Brock's thrown-together makeshift campfire. No tent. No facilities. No tent. Real live camping. Did I mention no tent? He brought a boom box, we play my tapes, he's laughing that I love *Black Sabbath*. We listen over and over, it reminds him of his teenage years too. He's got *Pearl Jam*. We talk the night away and hike around the next morning. We swim in the river, jumping off some long-forgotten rope swing.

Brock fills me in on deadly Australian things. There are snakes, box jellyfish, crocodiles, sharks, venomous spiders, even frogs, and best of all, stinging *trees*. Yep, leaves on certain types of trees will sting me, even if I just graze by, filling me with incredible pain so bad I'd have to vomit. *That's fantastic.*

"Yeah, when we watch your American movies where the kids are running through meadows barefooted, we can't believe it. We could never do that here. Our mums would kill us, plus we'd get stung or bitten a few times over," Brock chuckles, telling me of his childhood. Great, now we get to camp under the stars for another night. *I'm not creeped out at all... thanks.*

Airlie Beach

- Lucy in the Sky with Diamonds -
The Beatles

After heartfelt goodbyes and offers to stay in touch, I'm dropped off at the bus station and head south to Airlie Beach, a tiny tourist town with just a few street lights, lots of hostels, and a gorgeous harbor filled with boats going out to the Great Barrier Reef.

As soon as I get off the Greyhound, five people hold up signs that say "Stay at my hostel! Free breakfast! Swimming pool! Video games!" All for the bargain price of $20 Australian a night, and with the exchange rate, that means about $11 US for me.

"Hey, I need a job!" I yell out to no one.

"You could have my job," a sign girl says. "I'm leaving tomorrow. You just meet the buses for two days a week but get free room and board all week, too." Sounds great. I go to meet the boss.

The gray-haired, hostile-faced, hostel-boss lady, Janet, seems grumpy but says yes and shows me around. 15 bungalows surround the recreational area with a full kitchen, cafeteria style tables and chairs, and a large TV room with couches and bean bag chairs. An outdoor deck with a pool table and a swimming pool with loungers. The tropical garden surrounds the property with a pet baby kangaroo, called a joey, jumping about.

I'll stay in a bungalow with a few other workers. A few bunk beds, a microwave, mini fridge and table, couch, and loo. I meet my new roomies, Troy from Ireland, Karen, and Jules, both dark-haired British beauties, but Jules grew up in Abu Dhabi in the Middle East.

I call my mom to tell her where I'll be staying for a while. She tells me the double murder verdict for O.J. Simpson has just come in. After years of domestic violence, "someone" slit the throats of his ex-wife and her waiter friend who'd dropped by with her forgotten sunglasses.

I sigh, nodding my head. *Guilty, of course...*

"NOT GUILTY!" she tells me with the heavy, broken heart of a woman who works at the District Attorney's office and has witnessed the ultimate fail.

"But... *how can that be?*" I'm stunned.

She explains what it's been like for her office, where the case originated before it was transferred to downtown Los Angeles and what a media circus it's been. The key evidence that was not admitted in court, like a bloody doorknob on the outside fence *at O.J.'s house*, how O.J.'s bloody fingerprints were found on the light switch *inside O.J.'s home* containing DNA from his blood, and *both the victim's* blood. Even some in his car. Footprints in his size from rare Italian shoes in her bushes... shoes he said he didn't own but had been previously photographed wearing at a party.

Unless he had something to do with it, *why in the world would there be* ***any*** *evidence at O.J.'s home when the murder took place at Nicole's home?*

Mom tells me about the glove he'd pretended to try on, and how the liquid they'd used to test the blood can shrink the material. O.J. was not used to eating high-sodium jail food, plus he was switched from his usual type of prescription medication, ultimately leading to a swollen, puffy hand.

Mom tells me with DNA being a relatively new science in police forensics, how many jurors just didn't understand what 'DNA' even was. They confused it with having a common blood type. In a televised interview, one juror responded "Well there are millions of people that have Type A blood. How can they know who it's from?" *That's not the same thing, dummy.* And we both agree that such a well-known celebrity was never going to get a proper conviction anyway.

I tell my new flatmates about the verdict and they start in on me. "Well, 12 people found him innocent, so he's innocent." "I heard the cops framed him." "White cops trying to hurt black people."

Just because a jury found him not guilty doesn't mean he's innocent, it means they didn't prove his guilt. If a police officer frames someone in a death penalty case, the police officer himself gets the death penalty for himself, too. What about the two dead people? Maybe Los Angeles didn't want to pick up another billion dollars in damage on the taxpayer's tab for another race riot like just three years earlier, thanks to the Rodney King police verdict.

For once, *I bite my tongue.*

After a few days of swimming in the pool, getting to know everyone, even petting the baby kangaroo, I'm up for my new gig. I only drove once in South Africa, learning the "wrong side of the road" that time in the rain, otherwise Jay chauffeured me. Now I get to drive a minivan to go meet the buses. I meet my competition. Sean from Perth, a chubby, sandy-haired, completely loud, obnoxious type, and Kaeden, a short, muscular redhead with a beaming smile. He reminds me we actually met in the town of Cilicep on Java, months back while I was traveling with my Moroccan friend Hujaf. He also went to Bali, Darwin, and Cairns, a few weeks before me. The world just shrank a bit more.

The first bus pulls up, so I hold up my sign, smile, wave, and grab as many people as I can. I'll come back to meet the bus every two hours, but until then, I can swim, read, relax.

This is great. I probably pick up 15 people throughout the day. I'm fascinated with the different varieties of travelers. A Russian woman traveling with her three-year-old daughter. An old man who'd just lost his wife of 50 years and they'd always dreamed of seeing the world together. A couple of supermodel lesbians. And even a special needs couple. Mostly college adventurer party animals. I'm the new face of the hostel, so I'm suddenly friends with *everyone* staying there. No time to be shy as everyone wants me to tell them about Airlie Beach.

After my two-day gig, I pass the keys along to Jules. I'm by the pool with the supermodel lesbians and I hear grumpy hostel lady on speaker phone. Her friend with a tourist boat goes out on day trips to the Whitsunday Islands. And he needs a deckhand five days a week *if she knows anyone who's interested?*

I speak up and, visibly annoyed, she sends me down to the Abel Marina harbor to meet with him. I walk for about 15 minutes, and up ahead, I see a gray fin along the shore. My heart skips a beat. I get closer and it's a baby shark on the beach. A dead baby shark with its teeth torn out. I don't want to take a picture of it, it feels... *icky*... disrespectful. I leave it be and say a tiny prayer for him.

Walking onto the docks, I see some men waving to me from a yacht. I smile, wave, and climb aboard. The boat can hold 25 people, with open seating in a "U" shape along the back, and the captain's chairs under the covered rooftop. With a proper '*Gilligan's Island*' steering wheel, of course.

The owner, Davis, introduces himself. He's a sandy-haired, suntanned, happy guy in his 50's. His first mate Webber is a shorter guy with dark mop-on-top curly hair, probably in his 40's. Sherri's leaving the job soon and needs a replacement. They all invite me to stay to drink some wine and barbecue some fish they caught on today's run. Yay!

"Know anything about boats?" Davis asks.

"Not a lot, but I used to go water skiing and wakeboarding with my friends all throughout my teenage years," I hope that is a better answer than a flat-out NO. "I used to have to help with the ropes and cleaning up the boat after the big camping trips..."

"Great, you can start tomorrow. Beer or wine?" Davis says. *Wow, I thought this would be harder.* They ask me all about my life in America.

"We don't get many Americans through here, especially a girl traveling on her own," Davis comments. "You need a nickname... hmmmm. I think ALF. American Life Form."

"Hahahaha! I love it," Webber laughs. He throws me a boat logo shirt. I laugh, I can get into my new nickname. I just met these people and feel like we're long-lost pals. I'm going to love *boat life*.

Next morning, Sherri's training me all day. We greet the tourists, making them feel at home. They get seated, tucking away their towels and bags. I offer coffees and teas, candy bars, and trail mix. It's my responsibility to untie the ropes from the docks, bring in the attached buoys, jump aboard, tie up anything that might blow in the wind, and peel back the plastic windows to attach them to the interior. I'm jumping around the boat, copying Sherri, barefoot in the sunshine, about to visit White Haven Beach in the Whitsunday Islands.

This doesn't suck.

The yacht's been going for an hour and a half, now we're ducking in and out of sand bars, and *Voila!* White Haven Beach. The water's aquamarine, with sand so white it's blinding. The sea air is thick, the tropical forest right up to the edge of the sand. Heaven... White *Heaven.*

We putt-putt into the shallow dream water and Webber drops the anchor. We jump into action making sandwiches, handing out drinks. I make a fruit salad for everyone, just enjoying their chatter about how they all cannot believe the beauty of this corner of the world. After lunch, we all leave our clothes and climb down the ladder into the warm yet refreshing water.

"Time for a stingray walk!" Webber yells out to anyone who wants to come along. He grabs the hands of some of the children from the tour group and begins to skip into the pure white sand. The moment my toes hit the strand I sink in a bit more than usual. It's like a fresh powder snow day. *The sand is baby powder soft.* I hear subtle squeak sounds with every step.

"What's that squeak noise?" one of the young girls asks, so I don't have to.

"The sand is so fine it has a tiny bit of natural silicone in each grain," Webber informs all of us. He points to a family of stingrays near our ankles. He tells us they're docile and if we move slowly, they'll even let us pet them. We learn the *stingray shuffle*, gently shuffling our feet along the bottom to sort of kick them aside in case the rays are hiding under a fine layer of sand. *Good to know.*

This is where I'm going to spend five days a week? How did I get so *LUCKY*?

Sherri and I invite eight beginners for a snorkel lesson. We all kneel in the water, with us teaching them how to attach their masks and snorkels. Some are afraid to put their faces into the water, so I comfort them, giving encouragement. One Chinese lady bursts into tears—she's terrified and doesn't speak English. I take my time with her, showing her how easy it is to breathe through the tube, above the water first, how to squish the mask firmly enough to keep the water out, and how she can pop up her head in case she panics. No easy feat using nothing but charades. With a gentle nudge, she's in the water.

After a couple of hours at White Haven, we sail over to Hamilton Island and hit a snorkel cove where we can't touch the bottom. I sure hope my lessons have been enough for the beginners. Webber, Sherri, and I hand out breadcrumbs for everyone to throw into the water as we wait for the fish. As the schools arrive, we all climb into the whirlwind. Hundreds upon hundreds of fish, an underwater museum of sorts. Clown, angel, grouper, eel, giant clams with neon rainbows of algae hanging out of their mouths, sea urchins, bright fan coral, mountains of brain coral, an unbelievable prism of colors. The gentle motion of the ocean (*Yeah, I said that*) with sunlight dancing above our heads, I feel like I'm the girl with Kaleidoscope eyes.

My heart warms as I see that my Chinese friend loves it. She's the first in the water and the last to get out. Great success on my first day. I get a few pats on the back and a "Good job!" from the captain. Sherri's done training me, but I'm in my element. *And I get to come back tomorrow.*

Each day I feel more and more into my groove, meeting the other boat people, doing the ropes, greeting the clients, making the sandwiches, and teaching the lessons. Sometimes Webber even lets guests, me included, drive the boat for a bit.

Every group makes each day so different. Some days the groups are loud, drunk, and fun, while other days the groups are friendly families, and sometimes just a quiet group, the keep-to-themselves kinds of tourists. The ocean is similar. Sometimes a bright teal color with baby rippling waves, or a deep, dark blue with choppy waves, but Whitehaven is *always* the dreamiest, lightest aquamarine color, and as smooth as glass.

Some nights, I have fresh fish with the yacht captains. We enjoy fine wines and lots of laughs sitting on the boat in the docks. I even get a fancy yacht Club party night. Other nights, I'm invited out to pizza and beers with my bus driver pals and we exchange stories of our adventures. We even make a day to search for the hidden waterfall on the outskirts of town. After a beautiful hike, we find it sunk in between black rocks, flowing gently into a small grotto, like a cold refreshing private swimming pool.

Some of the other boat people invite me on their yacht as well, for *free*. Their catamaran is much more of a party boat, with beers and more beers all day long. A guy named Alistair and his crew insist I do a Resort Dive, which will be my first scuba dive ever. My first dive and I'm *buzzed*. Is this a good idea? I don't have to take any classes, just hold hands with Alistair and if I need anything, just squeeze his hand. They teach me the gear in five seconds flat and into the delicious drink I go.

I do the slow *Darth Vader* breathing like they taught me, enjoying the bubbles escaping slowly in the peace and quiet of the sea. Hundreds of fish simultaneously chomp on coral and it sounds like faint ticking. Even though I snorkel almost every day, scuba diving is something different, more adventurous, even, *sexier* somehow.

The day ends with a parasail off the back of the boat. Another first for me. It's dreamlike... I'm breathing under the sea, then flying high above it just moments later. The wind's in my hair while the captain keeps slowing enough for me and my parachute to leisurely drop so I can throw my head back to dip my hair in the water. Then back up. *This. Is. A. Great. Day.*

One day, I go for a long swim at White Haven and as I watch my hands gliding together through the clear water, parting gently, I think of my Dad and how happy I am that he's guided me here. I think of my family back home and wish I could explain the beauty of this daily profession. I've seen so much over these last few months, I've learned a thousand new things, I'm sure I've changed. I've still been meeting the most interesting people from all over the world. Their life stories become my knowledge of this world, and that knowledge transforms slowly into my new wisdom.

My flatmate, Jules, has the best stories from Abu Dhabi. She looks like a young Janice Dickenson and was a teenager during the first Persian Gulf War. All her British friends were terrified for her, "Shouldn't you get out of there? The news says it's a war zone and you're not safe!" She laughed and told them she'd been at the beach all day and might go to the mall later. She could hear a sonic boom off in the distance occasionally, but that was it. A Saudi Arabian Prince even took a liking to her and asked her to marry him, but her mother said "No!" like any good mother of a 14-year-old. *(Are you hearing this Priscilla Presley's Mom?)*

She asks about the highlife in Los Angeles, so I list off the movie stars I'd seen or met... while a local girl, with her jaw on the floor, says how absolutely dull her life is here. No princes or movie stars, just born and raised in Airlie Beach.

"But we're both here in your town looking for adventures!" I remind her. She smiles with gratitude. Jules says she'd love a job like mine and to keep my ears open for her.

The yacht clients share their interesting life stories as well. One chubby redhead in her 50's had lived in Papua New Guinea since she was 18. She moved there with her young husband to start a dairy farm for the natives. After a few weeks, all the cows were dead and eaten. The tribesmen couldn't grasp the thought of just keeping them for milk. After a few years of trying to replenish the cow supply and having them all eaten, she gave up and still lives on the abandoned dairy farm. She tells me of the tribesmen's violent, political riots every few years and how you just lock yourself in your home for a few weeks until it's over.

One day, a cute guy with bright gray eyes and light brown hair is on my tour. His name's Toby and he's only in town for a few days, on vacation with his father. At the end of the day, he asks me out on a date for tomorrow. I can trade a shift with Karen.

We decide on a nearby crocodile farm. As soon as we walk in, a show starts at a pond with a circular chain link fence surrounding a man with about 12 crocs. The guide is very *'Crocodile Dundee,'* rough and rugged.

"I am the *real 'Crocodile Dundee,'* he copied me!" the guide says arrogantly into his microphone. He explains what an expert he is amongst these docile creatures and how they respect him. I notice all the deep, jagged, purple scars on his legs. *An expert, clearly.*

He starts poking a croc with a stick, opens the creature's mouth, and puts the stick inside its jaws to prop it uncomfortably open. It seems like he doesn't respect the crocodiles, he's taunting them, bullying them. If I were a croc, I'd bite him as well. Toby and I share some laughs but move on to check out the grounds. Wild peacocks roam the gardens, and we even spot a rare albino. Toby seems nice enough, but the conversation flows a bit awkwardly.

We move to the koala cages, where two of the cutest little button-nosed climbers chomp away on eucalyptus leaves. The cages are really just massive wooden stalls with trees and toys inside, open to the sky. We're told we can hold one if we like. Yippie!

The guides tell us one's grumpy, like a bratty kid, and doesn't like to be held. They hand me his friend, the fuzziest, cuddliest, friendliest little guy. He grasps me around my torso, like a two-year-old toddler gripping onto his mommy. I'm in love. The little munchkin must weigh about 40 pounds, so he's heavier than I thought. He's looking into my eyes, feeling my hair and just keeps *smiling* at me. He smells like Eucalyptus and Australian red clay, mixed with a *hint* of wet dog. His long claws are daunting, and we're told the grumpy ones will attack with their claws in full force. Good thing I'm holding a *Care Bear*. I cuddle him for 20 minutes.

I'm not feeling a spark with Toby. This was a lovely day, but he and I can be friends. I say goodnight, give him a hug, and don't think much about it as he leaves for Cairns in the morning to accept a job offer. The next day I'm on bus duty and swimming all day. The following evening, after "work" on the boat, as Davis, Webber and I are barbecuing, up the dock walks Toby.

"Hey, Pam!" he calls out, waving.

"Oh, hey, Toby," I reply slowly, confused.

"Can you come and talk with me for a minute?" he asks.

"Sure..." I say, not really wanting to go, but dutifully stepping off the yacht.

We walk to a grassy spot next to the parking lot and sit to chat.

"Pam, I just drove eight hours to Cairns, and I realized as soon as I got there that I... I think I love you, so I turned around and drove the eight hours back. I never even went to accept the job." Toby blushes a bit, his eyes are slightly glassy, searching my face for the same feelings.

Well, this is... *awkward.*

I feel nothing but a little weirded-out now. We spent an afternoon together, actually two afternoons, but one I was at work and it was my obligation to be friendly, so that doesn't really count. So, one afternoon sight-seeing and cracking a few jokes. We didn't hold hands or kiss or anything. *Love? Kind of a strong word, don'tcha think?*

"Ummmm." I feel bad for all the trouble he's gone to. He's given up a job opportunity and driven 16 hours through the night and day. It could've been romantic. Except it *isn't*. It feels *stalker-ific.* "I really feel... like we... were just, um, getting to know each other for the day... so... I wouldn't be able to reciprocate your feelings... in this way..." I'm hoping this is enough of a gentle let down.

"But I drove all this way! I gave up a job!" *Oh, so it's not going to be enough.*

"I'm sorry you did that. I wouldn't have encouraged that even if I did have feelings for you," I say gently.

His eyes ablaze, he gets up, shouting, "I'm going to get a job on a boat like you and you're going to really fall in love with me!" He turns and jumps into a white van in the first parking spot. I stand while he starts the engine, and through the open window he calmly asks me to come closer. He says he wants to tell me something else. *Not on your life buddy.* You're trying to convince me you're not a stalker, and yet you're getting a job by mine and you drive a van? *Good for some quick kidnapping?*

I turn and speed walk back to my boat, while he's calling my name. Once I share this story with the two captains, Davis says to me, "Happens all the time, a fella falls in love but it's because of this job you've got on my boat!" he shakes his head as though he's seen this a hundred times.

Thanks, mate, just the job... not the Cindy Crawford eyes, DD boobs, and a sexy American accent? Yeah, guys hate that. Probably your boat after all.

Davis offers me a ride home and I take it. The very next day Toby's working on a boat three down from mine and he has to walk past me every single day. From now on. Morning and night. Oh, yippie.

After a few weeks of this life, with Toby fading into the background and never really speaking to me again, I get an early morning call from Davis at the front desk of the hostel.

"So sorry, Alf, you can't come in to work today," he says.

"Okay, no worries," I reply, not really upset as I haven't had a day off in ages. "See you tomorrow then?"

"Well, actually," he says, "You can't come back to work at all. I just got a call from some friends in town and they said Immigration is in town. Once in a while, they come through and do a sweep. They pretend to be customers for a few days, then whip out their badges and arrest you. You can lie and say you weren't working but they'll have you on video..."

Crushing news. "Oh, no! I love this job! Can I come back in, like, two weeks? Once things have calmed down?" I ask, realizing that means I can't work at the hostel anymore either.

"I'm sorry, darlin' they sometimes do the very same thing and come back a few weeks later too. It's about an $8,000 fine for you and a deportation back home, even if you have tickets elsewhere, but for me it's about a $25,000 fine plus I'd lose my business!" he confirms. *Wow, I had no idea.*

"And that call I got went all through town, so everyone's panicking right now, I'm sure," he continues.

"Well, hey, Davis, this has been a dream come true for me, so thank you from the bottom of my heart. You all made me feel like an old friend, like a local. I'll never forget you or this place!"

"Aw, thanks mate. Know anyone looking for a job?" he says.

"Actually, I might! She's a beauty from Abu Dhabi, and she has the proper paperwork. I'll ask her now," I hope to help two friends.

"If she can start today, you can come on board and have a last day as a tourist today and I can train her if she's interested," he sounds hopeful.

"Okay, hold on!" I shout and run off to find Jules.

Of course she says yes, and we walk over to the marina together. She loves it. And just like that, she's in and I'm out.

I pack my bag and say my goodbyes. One of my clients from the boat a while back, Stuart, had offered me a place to stay if I ever got to Hervey Bay down south. Sounds good enough to me.

Hervey Bay

- Enter Sandman -
Metallica

My friend Stuart's flying out tonight for a business trip for a week, he apologizes for the poor timing. But asks if I'd like to join him at his Surf Club right now. As a matter of fact, I do. He can drop me at a hostel afterwards, on his way to the airport.

He introduces me to a tall, muscular, red-haired, fully freckled man named Hamish, and his girlfriend Sadie, a much younger sultry lady with blue eyes and a sleek jet-black bob. Once they hear I have nowhere to stay, they offer me an extra room at their home. At this point, I'm feeling the flow of the universe and don't even question it.

I happily place my bag next to the mattress on the floor of the downstairs room of their home, while Sadie hunts for some fresh sheets and towels for me. There's an exercise bike and a few dumbbells. The small, attached bathroom has a broken window, as though a baseball had been thrown through it, with a gorgeous plumeria tree right outside. They've invited people over for a barbecue tomorrow night and I'm included as well. It happens to be my 25th birthday tomorrow, so it sure would be nice to have some fun with new friends. I meet the 15-year-old son, Adrian, a surfy looking guy, and young daughter Minnie.

I'm exhausted after all the traveling down by bus and an afternoon spent drinking beers and meeting new friends. I hit the sack early. In the middle of the night, nature calls so I head into the bathroom, turn on the light and *there it is.* A Huntsman Spider. At face level. In my face. Bigger than my hand. They're the only non-venomous thing here, but they're *fast* and *furry* and they *jump*. I'm wide awake.

I don't know what to do. I can't kill it, I sure as hell can't go back to sleep, *obviously*. I decide to wake up Adrian. Poor guy, I knock on his bedroom door and he's glad to help. He grabs a broom to shoo it out the hole in the window.

I go back to sleep, warily, only to hear a knock on my door. *Adrian again? Maybe he's worried it came back through the window?*

I turn on the light, "Yes?" I say groggily.

The door creaks open and there's Hamish. Naked.

"Just wonderin' if you need an extra pillow?" He smiles. The creepy smile isn't touching his eyes.

This. Is. Not. Good.

"Ummmm, no, I'm good, thanks..." I say, hoping that'll do it.

He enters the room anyway. My heart pounds in my eardrums. *Really dude? Your girlfriend is right upstairs. Your* ***children*** *are right upstairs. Where are those dumbbells?*

"Well, I think you should have another pillow," and smiling, he climbs into bed with me.

Naked. Oh. Shit.

"Oh, shit! I'm totally sorry if I led you to believe I was that type of girl, but I'm not. Definitely not," I say smiling slightly, but speaking firmly.

"I was nervous coming down to see you, but I was up there in bed with Sadie and couldn't stop thinking about you..." he keeps trying.

I smile, say nothing but pull the sheets up a little closer to my neck.

He sighs. He smiles. He gets up and walks his pasty-white-freckle-butt to the door. He grabs the knob and looks back at me, hopeful.

I blink, reach for the light, and turn it off. He closes the door behind him.

Thank God.

I remember feeling no thoughts of personal safety in the jungles of Sumatra or on the edge of the volcano in Java, but the nightclubs in Bali filled with drunk tourists reeked of *rape-y* possibilities.

I'm 25 today!

I'm a quarter of a century. Life is going well. Well, interestingly anyway. Yes, I'm traveling around the world on my own terms, but everything I owned had been stolen, I have no job, no home, and no car. No boyfriend, no husband, and no kids. Mine's not the dream picture of where I thought I'd be, but in just a couple of short years, my dreams have changed dramatically from what I used to think of as "success."

As I venture out into the kitchen, wondering if I should leave after last night, I see Sadie and she smiles.

"Brekkie?" she asks.

I nod as she scoops some eggs and fruit onto my plate. As I eat, she invites me out to lunch.

"A little birdie told me it's your birthday today!" she's excited to be pals.

"How in the world could you know that?" I exclaim.

"I knew you'd wonder! Hahahaha!" she giggles. "Stuart said when you met on your boat, you'd all been talking about birthdays and he remembers everything. He mentioned it to us at the Surf Club, that's why we were so happy to host you at the barbie," she winks.

I can't believe it. I'm already treated better here than by some of my friends on my past birthdays.

She and I decide to go to the beach and have lunch before she has to pick up the kids from school.

The sun's shining, the water's a gorgeous royal blue and the waves are small but perfect. Sadie explains how Fraser Island blocks some of the big waves. "It's a small island made of nothing but sand, where people can off-road. It's so fun over there! Can be a bit scary, though. Sometimes sharks come up out of the water and beach themselves just for a nibble of any human standing too close on the shore."

"What?" I can't believe it. "You mean there are shark attacks on swimmers over there?"

She giggles. "No, I mean there are attacks if you're standing in your trainers up on the sand near the waterline, *not even in the water!*"

WHAT?! For some reason I picture the old 60's '*Batman*' TV show where Batman's hanging from a helicopter ladder, punching a shark in the face.

We lay out our towels and Sadie strips off her bikini top. Oh yeah, I forgot. Topless beaches again.

"So, I heard what happened last night..." she says.

Oh no. "Oh yeah?" I say, trying to be cool. What could Hamish have said? He was at work when I got up. Is it a lie? Probably. Most men who get rejected will say whatever makes them feel better.

"No need to be afraid of a huntsman. They aren't poisonous!" she laughs.

Whew. *By the way, did you know your boyfriend is a cheating asshole?* "Oh, good to know, it was bigger than a Saint Bernard!" I make light.

We talk all day about her relationship with Hamish, how much she loves him, but he hasn't proposed yet. I don't mention The Scandal. At an outdoor café, she pops open champagne to celebrate my birthday. What a wonderful new friend.

"You know, I was a bit jealous of you. The way Hamish looked at you right before he invited you to stay with us..." she says, a bit of a buzz on now.

"Oh, no worries there, I'd feel the same way if you were staying with me and my boyfriend back home!" I'm buzzed as well, trying to comfort her.

"Oh, you've got a boyfriend back home?" she's hopeful.

Yep, I do now. "Well, we're on a break while I wasn't sure how long I'd be away, but we're still very much in love," I say. Is this a white lie? Like, great haircut? Your butt looks great in those jeans? We giggle about boys the rest of the day.

I help chop the salad for the early afternoon barbecue and enjoy meeting new friends. Hamish acts like nothing happened. It's a lovely birthday. That night, after the guests have left, we settle in to watch a new educational show. About sex. Their teenager is still with us and Hamish is getting excited. It's the first night it's airing, and it shows close-ups of hard-ons, spread legs, and what to rub and where. Now, I'm not exactly a prude, but this is a bit much. At 8:00 p.m. on regular TV. Yes, it's "educational" but seriously, not 10:00 p.m. on HBO? It's awkward watching with this family, I can tell you that. I head straight to bed.

I've been in bed for a while and nature's calling again. I get up, turn on the light, and there is a Huntsman Spider. At face level. But this time it has only three legs. Two on one side and one on the other. And it's crawling slowly. In a bit of a circle. It's almost too *funny* to kill. Except it's *not*. I go wake up Adrian again. He comes down and does the same broom trick. I'm extra creeped out with the three-leg thing. He insists it's probably a neighbor kid that trapped it and pulled some legs off. *So Gross.*

Back to bed and, déjà vu, there's another knock at the door. I turn on the light and yep, naked Hamish. Now I'm pissed off. I should've left today. I just wanted to be Not Alone for my birthday. Was that too much to ask? I guess so. Here comes another creepy vibe.

He says nothing and slips into the bed with me. He's drunk. No longer angry, I'm terrified.

He tries to kiss me, and I struggle away. Thinking fast, I pretend I'm weighing my feelings. "I'm just at a place where I'm finally realizing that I actually might be into... girls. Sadie was topless today, and I thought she looked so incredible. This entire trip I'm discovering new sides of myself..." I turn away, praying he buys it.

"I can go wake her up right now!" he's excited.

"I think another night to re-evaluate my heart..." I leave it hanging in the wind.

"Okay, but tomorrow night is the night!" he growls as he runs his hand over my waist and hip before he jumps up and heads for the door. Again.

I cannot sleep, obviously. Dawn is breaking, so I quietly and quickly pack my bag, leave a goodbye note to Sadie and the kids, stealing a sandwich as a goodbye to Hamish. I walk out the back slider and out the gate. I walk toward the beach, sit on a bench for an hour, and eventually find a taxi that takes me to the bus station and out to Brisbane.

What a weird little stop in my travels.

At least I'll never have to see that Circus Clown Crotch again.

Brisbane

- Boombastic -
Shaggy

I find a hostel and meet a tall, skinny German beauty with long black hair, Mirra, who wants to get hammered at a nightclub. I go out dancing with her but, honestly, I'm exhausted and she's putting shots away like her ship's going down. She's angry everyone hates Germans because of WWII.

Hmmm, what about WWI?

Brisbane's pouring and boring for days. Mirra is still angry and drunk. I miss Airlie Beach. I miss Bali. I miss jumping off the pier in Durban. I wonder what my new friends are doing. Camryn's surfing. Maday on another volcano hike today. Dita loving her kitten. Lucy back home in England.

I buy a one-way ticket to Auckland, New Zealand. I still need a visa from the New Zealand Embassy, but once there, they want to see my ticket *out* of New Zealand, even if I'm not sure how long I'm staying.

"Well, I can't give you the visa if I don't see an outgoing ticket," the lady clerk says, raising a perfectly drawn eyebrow.

"So, I could go get the high priced, fully refundable ticket? Then get a full refund right *after* I get the visa?" I smile with a shrug.

She lets out a long sigh. She shrugs too, fully annoyed that I know that. I say I'm just going to get out of Australia because my Oz visa is running out and I'll stay for maybe two weeks. She mentions I'll love it and hands me my passport back with the completed New Zealand visa included.

Good-bye Australia. Fantastic memories, mostly. Hello Kiwis!

New Zealand

North Island

Auckland

- Stuck in the Middle with You -
Steeler's Wheel

Auckland sits in a narrow passageway in the north of the North Island. It's surrounded by ocean breezes. As I step onto the tarmac, I feel the cooler, crisper air. Every Kiwi I've ever met has had the best sense of humor and speaks lovingly about their picturesque country. And bonus points, there's nothing venomous at all.

Through the taxi window, the city looks very clean, with lots of book stores and international restaurants. The cross walk makes all directions of cars stop at once so the crowded sidewalks can allow the pedestrians to cross to each corner, but also walk like an X through the middle of the busy highway if need be. I like it.

In the hostel rec room, everyone seems a bit more studious, reading books, no one's talking. One girl is making homemade vegetable soup. Another is doing yoga poses in the garden. Very different from the Oz party vibe. The check-in girl tells me about the cool towns to the north, up to the Bay of Islands, ninety-mile beach, and one called Paihia, with tourist boats and lots to see. I book a bus ticket right away. Maybe I can get another boat job?

Once in the small town, I head over to the marina. All the brochures say the "Hole in the Rock" is the thing to see. The crew tells me they *are* looking for a deck hand if I want to come out for a try with them today? Yes, please!

Out on the water, the wind is freezing, and the water is very, very choppy. It's lovely, but I think it'll be hard to try to recapture my Whitsunday Island life. I turn down the job, thank everyone for the day, head back to Auckland to find my *New Zealand* adventures. I realize I got the day trip for free just by applying for the job. *Must remember that trick.* The entire country is about the size of California, so I might just sightsee and then head back to Australia, like I told the Embassy Lady.

A few people mention that Rotorua has white water rafting, a few hours south. Sounds good to me, so I grab my bag and head down that day.

Rotorua

~ Hey You ~
Pink Floyd

As soon as I step off the bus, the rotten-egg stench of sulfur is thick in the air. It's a tiny town with a one-block main street surrounded by suburbs and farms. I head straight to the white-water raft place and speak with the brown-eyed, 20-something tour guide, Oliver. There's room for me in 10 minutes. I'm told there's a seven-meter waterfall and sometimes people flip out of the raft. If that does happen, just grab the nearest rock, and wait to be picked up because there's an even larger waterfall a bit farther along.

Seven meters, huh? That must be like seven feet, right? No biggie.

All of us tourists get into our swimmers and wetsuits. *Why does the fattest, hairiest guy insist on wearing a tiny Speedo?* They hand us oars and helmets while we shove off. The guides crack jokes along the way, teaching us how our weight is needed to balance the boat. Oliver might yell out for us to stay up on the rim of the boat, or to jump back down onto our butts inside the boat. He'll yell out when to row and when to hold back.

The gentle rolling hills along the riverbanks are a stunning, neon green. I feel like I could be in Ireland. And did I mention the sheep? We float past ranches with hundreds of sheep. I learn there are *way* more sheep than people in this country.

We're all laughing and rowing, jumping up, then down, laughing some more. My new friends are from Czech Republic and Norway. Everyone's struggling with the thick Kiwi accents of our guides and I'm somehow the translator, which brings out belly laughs. The water gets faster as we steer around large boulders. Only whitewater rapids now, as we get into our groove as a team. Up, left, paddle, down, paddle, right. Icy water sprays my face. I'm giddy enjoying this newest challenge.

"Here comes the waterfall!" the other guide yells out. "Everyone, lean back as far as you can, and hold on tight to your paddle!"

The loud thundering roar seems way too ominous for seven feet.

With that, I find out seven meters is Seven Meters, which is really *21 feet*.

My eyes bulge out and my stomach drops as we go over, screaming and laughing, falling, more screaming, more laughing. We land roughly with a bounce in the whirl of the white water at the base of the river.

Disappointment overwhelms me that we didn't flip over. It's like the 1994 Northridge earthquake in Los Angeles last year. I was in San Diego at the time and *felt left out* that my friends and family had gone through something so exciting.

There must be something wrong with me.

We go on through the mild currents, winding gently through rocky cliffs, mossy gardens, ending at a cove. Oliver asks if I want to go another time for free, because they need more bodies for the next group. Why, of course, kind sir! I end up eating sandwiches with the guides as we wait for the next tour. They ask me all about America, and I fill them in on how long I've been traveling. When I ask about the thick sulfur smell, in unison, they shout "What smell?" It seems I've stumbled onto a local's joke.

I'm in the raft again, introducing myself to the new group. They must think I work here, it's just a habit after the yacht job. I'm ready and willing, heart pumping with excitement to go a second time. I'm fully prepared for the waterfall. It's another beautiful river dance and this time, over the falls, instead of a bounce at the bottom, we do *flip over!*

The icy water engulfs me and I'm holding my breath. I pop up under the dark, overturned raft. I've got to hold my breath again to swim out because we're still moving rapidly through the current. As I get sucked along, I've missed the end of the rope attached to the raft. My feet, knees, shins, tailbone, and elbows get bumped against the immovable boulders, bruising me in too many areas at once. I've mistakenly thought the stream was mild, but now that I'm trying to swim it, it's rough.

This isn't fun. This is dangerous. *Okay, a little fun.*

I swim to the edge while the water's jostling me along. The first rock slips through my fingers—I'm not quick enough. My elbow bangs a rock sharply. I gulp water and cough a bit. A couple of big kicks and I reach for a boulder. Again, I can't find a grip on the slippery moss in the fraction of a second that I have to hold on. Okay, now I'm worried. Again and again I try, and the fifth time is the charm. A boulder with the right handhold to hook into, with enough room to brace myself against the current.

I hang on for dear life as my feet are swept out from under me along the tough tides. I'd floated around a corner and am alone in the river for about 15 minutes, wondering if I should do something—anything—differently. My thoughts go to LA and my family back home. I called my mom and let her know I was in New Zealand, right? Oh, yes, once I'd landed. Do my new friends know I floated off? Because I was under the boat a few minutes, I'm not sure if everyone's ahead of me or behind me? Are they all okay?

"Ahoy, there, Matey!" calls out Oliver. I look up and see the boat with all my new friends in it. Oh, it probably took a few minutes to flip the boat over and collect everyone else... *whew.* My heart's pounding, my adrenaline's rushing. Everyone grabs my elbows and yanks me in, while I flop face first like a caught fish, exactly like the graceful debutante I pretend to be.

Back at the truck, we're offered a snack of "bikkies" and hot tea. When Oliver hears I have nowhere to stay, he invites me to stay at his home with his new wife for a few days. Yes again!

He calls ahead and by the time we arrive to a lovely three-bedroom home with solid wood floors, his wife Simone welcomes me with a warm hug hello. She gives me the tour with the smell of a hearty lamb stew simmering on the stove. The guest bedroom is filled with sewing machines and knitting yarn. In the five days since I'd arrived in New Zealand, I've been so *very cold...* my blood is still accustomed to the much more tropical weather of Queensland, Australia.

Simone must read minds, as she says warmly, "Why not go grab a hot shower after your chilly day today, then supper will be ready!"

Sounds like Heaven to me, Snow White.

All warm and cozy now, we're laughing and getting to know each other. They ask about America, Hollywood, even my childhood and what my Christmas mornings were like as a child, my schools, even what my boyfriends have been like. They explain Rotorua has the nickname of "Roto-Vegas" due to the flashing Christmas lights above the one liquor store on Main Street. They tell me all about their wedding and Simone brings out her wedding dress, veil, and tiara.

"I made them myself," she says modestly. My mouth drops open in awe of someone who can create such beauty. American wedding dresses lately are of the fluffy, ballooning, ballroom kind. This is svelte, sleek. A spaghetti strap satin dress with a slight train, with crystals and beading all over the bust, shimmering down like ombre raindrops in all the right places. I LOVE it. The veil and tiara are perfect complements to such exquisite work.

"Oh, God, this is *so elegant!*" I can't help but gush. I find out this is her first dress and she's trying to start a wedding dress design business.

"Well, you have found your calling! It looks like a real dress!" I say.

"It <u>is</u> a 'real' dress!" she snaps angrily and frowns, yanking the veil from me.

Uh-oh. Foot in my mouth? I quickly apologize and mumble how I meant *'professional.'* I make a quick excuse to head off to bed.

I stay up for a while, making use of the sewing equipment to mend my sloppy, fraying, stretched out Only Bra. I contemplate that this bra had cost me only about $5 when I found it on sale. I'm sewing my $5 bra. What kind of life have I chosen exactly? It used to be cream colored but now it's the non-dazzling, non-elegant color "Backpacker Beige." Beige is normally a lovely color, but this means it's been severely sweated in. Washed roughly in the wrong grungy soaps. Grimy bus rides, shoddy train rides. Goats. Chickens. The beaches and desert of South Africa, the thick humid jungles of Indonesia, the dusty outback of Australia.

But, being chased by a herd of elephants, climbing a live volcano, scuba diving, and parasailing has been a pretty good tradeoff. I stitch my worn-out, used-to-be-royal-blue, saggy bikini as well.

The next morning Simone tells me about Lake Taupo, a fun backpacker town about an hour south. She still seems cooler to me now, so instead of staying for a couple of days as I'd planned, I decide to move on.

I've hurt her, and I hadn't meant to. My California slang had led me astray. Constantly being the "guest" can be hard sometimes.

Lake Taupo

- Love Shack -
The B-52s

Taupo's a beautiful small town along the edge of a large lake surrounded by trees and rocky cliffs, with a snow-capped volcano on the far end. Kind of reminds me of Lake Tahoe back home. At a backpacker hostel, I grab a bed in a room with seven bunk beds. 14 people. Probably drunk. Ugh.

It's a gorgeous sunny day so I change into my bikini, my Bali blue sarong with suns on it, and a white T-shirt. My hair's up in a bun and I have no makeup on. I stroll down to the lake for a relaxing afternoon with '*Weaveworld*,' another book by Clive Barker. Awesome storyline, about how a string from a magic carpet falls off, and a few magical creatures fall out—for the first time in the real world.

I pass restaurants and a couple of pubs along the main area. I order some fish and chips from the Fish and Chip shop. Kiwi accents sound like "Fush und Chups." I buy a soda that I've never heard of called an L&P, Lemon and Pairoa. Still wary of my Singapore soda with the floating nuggets, I'm cautious as I try my first sip. It tastes like my mom's British secret stash of Lemon Curd on toast. Sweet, lemony, sugary yumness. And sweet potato fries. Excellent choice.

I bring it all down the wooden steps to the beach. Along the lake a few people lie on their beach towels, smile, and say hello. My toes have to feel the water. It's deliciously cool, but not freezing. It's summertime here, almost Christmas. It would be nice to celebrate the holidays with some new friends. I spend hours soaking up the sun and reading, watching the birds fly around and the breezes gently nudging the sail boats along.

I know this place feels more like home, like Airlie Beach did. I'm going to look for a job and see if I can stay here. I walk the few blocks to the hostel, feeling all relaxed and tanned. As I walk into the group kitchen, I hear some men talking about where to find some girls.

"Well, look, here comes one now!" A deep British accent attached to a guy with the blue eyes, straight black hair down his back, square jaw, and some tattoo sleeves.

"Well, maybe you should introduce yourself!" says the tall, lanky, sandy-haired man with the accent from Northern Europe somewhere.

"Well, maybe *I* should introduce *myself!*" says the strawberry blonde chubby guy. I couldn't even guess at his accent.

I smile and flirt.

"Hello, I'm Pam!" I bat my eyelashes, forgetting that I'm bare faced.

All three men jump up from the kitchen table and introduce themselves, Sebastian from England with the black hair, Sven from Denmark with the sandy hair, and Logi, the ginger from Iceland, of all places. They'd seen me check in earlier and had been trying to guess where I was from. Italian, Brazilian, and Persian made the list. No one thought of America until they heard me speak. They're all in their late 20's and have been traveling around the world working along the way, like me. Sebastian works at the pub called Crazy Horse that I'd spotted down by the lake. He thinks the only other bar in town needs help.

"You should stop by there after you come out to dinner with me. With us. With me..." Sebastian says.

I agree, happy to meet some traveling locals.

After getting dolled up, I head back into the kitchen to some wolf whistles. *I needed this.* All the men are dressed nicer as well. Logi must leave for work as a waiter around the corner and so Sebastian, Sven, and I pop out to the... Fish and Chips place. Oh well, it was pretty yummy earlier. I order the same thing I just had for lunch, from the same guy. He doesn't recognize me with my fluffy, curly hair midway down my back and my face all done up. I pay for myself as I notice Sebastian doesn't offer.

Sebastian is a flirt though. He asks all about America because he's never been there before. Sven is much quieter, more serious. After dinner Sebastian grabs my hand to hold it as we walk. It feels *weird.* I haven't had any affection for absolutely *ages.* He leads me upstairs into the bar, The Holy Cow, and introduces me to Mike. He's the owner with shaggy brown hair, a little soul patch, and hazel eyes.

"Yeah, we need a hand, can you start tomorrow night?" Mike asks.

I guess I'm staying here! Yippie!

I look around the bar, my new work away from work. Wooden floors, a few pool tables, wide bay windows to watch the sunset by the lake.

"Want a beer now? And here's your T-shirt, then," Mike hands me a green T-shirt that unfolds to show a cartoon cow with angel wings and a halo over her horns, flashing her udder. *I'm going to like it here.*

Sebastian has a beer with me, and we chit-chat with the other patrons. He knows everyone in town. He's off to work, so I stay to make my own new pals. Simon's the bartender. He's also Mike's brother, and he has light brown curly hair, shiny blue eyes, and a goatee. Did I mention his square jawline and dimples? *Yeah, I hadn't noticed either.*

Louie's a blond Dutch guy, standing at 6'7" wearing a cow-horn headband, and he's the co-owner. Victoria, with long deep red hair, is playing DJ behind the bar. Walter's a big, very wide, very strong Māori native who would seem intimidating as the bouncer, but who greets me with a warm hug and a smile. Everyone's outgoing and friendly, and right at 9:30 p.m. the place starts filling up fast.

In walk the Kiwi Experience bus drivers, Bennie, and Timmy, with their bus loads of backpacking 20-somethings, touring the country. A different driver pops into the bar every few days, but every three weeks, each circle back around with a new group of tourists. Then in walk the three DJ guys from the local radio station—Jimmy with the blond mullet, Mick with the black mullet, and Paul, the most ruggedly handsome stud that I've ever seen in my entire life. I blush immediately as I meet him and nearly faint from seeing his piercing blue eyes. *(*Thank you, Paul, for helping me remember the Taupo memories, and yes, I'll write your exact description of yourself)* all from the 'FISH FM' radio station.

But wait there's more! Next in walk the bungee jump guys and the skydiving crew. And each of them brings in large groups of people they had jumped with earlier in the day. My new bosses, Mike, Louie, and Simon make sure to introduce me to all.

I know! It's a blur.

I'm trying not to drink more than a beer even though everyone's partying already. *I'm still at a job interview.* I meet Michaela, a cute Nicole Kidman look-a-like. We crack jokes and I feel like we're old pals. Simon's training me, showing me how I'll pull beers as a bartender and to collect the empty glasses for washing. John from Sweden is the cook, sporting a brown ponytail. His girlfriend, Linnea, smiles with dimples, her light brown hair with bangs down to her waist, and she's doing cartwheels around the bar in a tiny skirt. She teaches me to say "*This tastes like shit!*" in Swedish at every dish her boyfriend John has cooked.

After midnight, the B-52's *'The Love Shack'* song comes on and I'm pulled up onto a table by 10 people to do a hilarious dance I know nothing about. No one's bartending. No one's collecting glasses. The entire bar is doing this dance with us. I'm just following along as best I can, laughing all the way, feeling like a local already.

I decide to leave after the dance because some of my new friends have told me to go get a tax ID number for New Zealand, and I should get it in the morning from the bank. I head back to the hostel and climb into my bed while the room's empty.

I'm awakened at 5:00 a.m. by my other 14 roommates. I guess the bars have closed. Drunken loud people yelling, turning on lights, fucking, vomiting. *Lovely*. No one notices me, or cares. The smell of booze and cigarettes is so strong I can't sleep. I get up to watch the sunrise with a cup of tea.

The owner of the hostel, a short, ruddy-faced man named Angus, introduces himself. We talk about how I just got a job and will be living here for a while. He offers the couch in his office for $5 a night so I can get some real sleep instead of bunking with the drunks. Very nice!

He also shares with me that the bank in town can't grant me an ID number and that I'll have to travel back up just north of Roto-Vegas to the headquarters. His friend is driving up to Auckland in 10 minutes and can drop me off there, no problem. Thank you again!

A few hours later I'm done with my chore and in need of a ride back. I get a wild hair and stick out my thumb on the side of the road. I wait, I wait, and I wait some more. I'm walking south and sticking out my thumb, but no cars are driving by at all.

Two hours later, and turning down a single slow tractor, *not kidding*, a low-rider-type of El Camino pulls up with three fully tattoo-faced Māori men. They're heading to Taupo as well, so I jump in and off we go. After waiting for hours, I would have taken a ride with Charlie Manson himself.

I might as well ask about the tattoos... I'm new in town, they'll understand that I'm just a dumb tourist, right? They kindly explain that when tats are on someone's face it represents the tribe of your mother on one side, and the tribe of your father on the other side. Lots of women do the smaller version and get them on their chin or inside their lower lip instead. The full body tattoos are done with the story of your life, childhood on one side and adulthood on the other.

They're also honest with me that it's more common *in prison*, and that they've each been in for different reasons in their lives. *Gulp.*

"See? I can relate. I'm one of you guys!" I show them the tiny paw print I got on my foot the night before I left for Africa. They laugh at me for ages. It *is* pretty puny.

I learn the Māori clans were not the first people to inhabit the New Zealand islands, the Moriori tribes were here first. Every last human in the tribes was ***eaten*** by the Māoris once they arrived from the surrounding islands.

Fantastic. So, I guess I'm hitchhiking with Tattoo-Faced-Cannibal-Descendant-Prison-Parolee-Random-Strangers. *Just a regular Tuesday.* Did I mention they reek of weed? The shotguns in the back of the seats? No? Ah, well, not important now. They invite me to a barbecue at their home as we pull into Taupo, but I've got to go to my first night at work, so I regretfully decline. *I'm not exactly sure whose ribs would be on the grill, after all.*

At the hostel, I move my bag into the office and get changed into my Holy Cow T-shirt and little black short shorts. My new uniform. I like it.

"Hey, sexy lady!" The door shoves open and in walks Angus. Without knocking. This could've been awkward 10 seconds earlier. He gives me a hug for some reason and kisses me on the cheek. I immediately regret moving my things into the office. I haven't thought this through.

I head to the rec room and in walk Sebastian, Logi, and Sven.

"Hey, Pam, so you got the job! Cool!" Logi says grinning.

"Very cool," Sebastian says, eying my tanned legs.

I explain how I've just moved my bags to stay in the office, but now Angus has made me extremely uncomfortable and I wonder if he's going to burst in on me while I'm sleeping.

"The three of us share the next room, just us. You can join us. We're all on the same schedule as well," Sebastian says, while the others agree.

"You're sure?" I'm grateful and grab my bag. They even have their own bathroom. I feel way more comfortable already. We all head out to work. Sebastian holds my hand as we walk and gives me a long kiss as we part ways. Yes, Taupo's going to be great.

The Cow is crazy from the get-go. The bungee jump guys are there with their bungee cords as the crowds fill up the place. "Toad" is the funniest bungee crew guy and invites me out for a jump while I'm here in town. They've organized a race between two guys, each wearing a cord around his waist, and despite the resistance, they're to run the length of the bar and reach a beer in between the legs of a girl sitting in a chair. *Yeah, **I'm** the beer girl.*

I'm holding my knees around the cold can while the crowd cheers on. Each guy is slowly getting closer until they're grasping with their fingertips until—*SLAM*—one guy can't hold it any longer and snaps back, falling to the floor, laughing. I scooch forward a millimeter so the second guy can grab the beer before he, too, goes hurtling back, shaking the can and spraying everyone.

I meet the skydiving crew. J.R., who's a tall Sean Cassidy twin nicknamed because of the old TV show '*Dallas*' with the iconic question "*Who shot J.R.?*" Rien, the shorter, more serious, brown-haired guy with the chiseled features. The blond and blue teddy bear Trev, whose real name has an unfortunate pronunciation that sounds like "Urine." So, Mike and Simon made up Trev and it stuck. They're all from Holland. Their boss is Van, a good-looking older guy with a gray mullet. He owns the skydive field. They invite me for a jump while I'm here.

I'd always been curious about extreme sports, so I tell them no guarantees, but I'll think about it.

Part of my job is telling tourists how much fun this town is, tell them to bungee and skydive, to horse trek, rent a boat, and check out some of the cool restaurants in town. In turn, all those businesses bring their people in to The Holy Cow.

Every time I pick up a food order, I tell John it "*tastes like shit*," and he laughs and laughs. The music's pumping and I get into a groove for hours collecting empty glasses, then switching to be a bartender while Simon or Victoria walks around collecting. Everyone ordering at the bar just shouts "Beer!" without giving me a brand. *I'll choose whichever one I want for you... so weird.*

The party stays strong until 4:00 a.m. when we sweep up and wipe down tables covered with the footprints of all the people who danced on them. I mean from *us* when the song '*Love Shack*' comes on again.

I walk the one block home and climb into bed, and the boys all head home around the same time. The sign by the light switch has the saying "Turn out the fucking lights!" in eleven languages. Sebastian climbs into bed with me, but just spoons me because we're both exhausted and fall right asleep.

Mid-afternoon, Sebastian climbs out of bed, naked, and I stare. I frown. I stare again. I must have a look on my face like I'm trying to do calculus in my head. *What's wrong with his dick? It looks like a frightened turtle! Is that contagious? Is he a burn victim? What the hell?*

"Good morning, Miss New-in-Taupo. Just let me put in my contact lenses and we can make some breakfast, maybe hit the lake?" he says. *Awesome, he couldn't see my face yet.* I realize it must be the *First Uncircumcised Penis* I've ever seen, and I try to act normal immediately.

"Sure, sounds great," I say in unison with Sven and Logi. Angus sees me come out of their room, his red and ruddy face contorts from shock into anger.

"Thanks for the offer, but I think I'll be more comfortable in here," I say.

"Well, it's regular price then!" he barks at me.

"Okay," I say. We walk out quickly. *Whack-o.*

We spend the afternoon at the lake eating, napping, swimming.

"I think I want a new tattoo on my ass, like MOM, or WOW, you know, around the hole... but something way funnier. Not sure yet," Sebastian tells us, strumming his guitar to Pink Floyd.

"Why not get Bugs Bunny's mailbox on one cheek, with rabbit footprints walking into his bunnyhole? Now that's funny!" I say, my mouth full of turkey sandwich. Not sure where I dreamt that one up.

A look of pure delight crosses his face as he puts down his guitar and turns toward me.

"That's IT!" he shouts and grabs my shoulders, shaking me with enthusiasm. He takes out his sketch pad and starts drawing what it might look like. "That's the one I've been waiting for! I'm getting it today!"

There's a tattoo parlor right next to the Crazy Horse and they can do it by worktime, but only if he starts right away. Hours later, I stop into the bar to see how it went, and there's Sebastian already with his jeans around his knees as he shows all his pals the new art. He sees me and points me out, introducing me. He tells them I'm the one who thought this up and everyone likes my sense of humor. I recognize some of the Holy Cow friends and even the tattoo faced guys I'd hitched with. We all catch up before I'm off to work.

*****

Things go on like this for a few weeks, hanging out with the boys at the lake some afternoons, other afternoons are spent with hilarious friendships forming at the Cow. Every single night the skydivers come by, each with new chicks that had gone for a jump with them that day. The FISH DJs need me for a cleaning lady. Paul- the sexiest, most gorgeous hunk of burning love *(*as per your request, Paul)*- Mick, and Jimmy's house is just down the street. I'm invited to the station to do some voiceover work. They love my American accent and have me run some ads. I get the nickname La-La. If they play a long song, like a 22 minute Led Zeppelin '*Stairway to Heaven*,' they've admitted it's just to sneak home to take a crap. *Fantastic information I did not need to know.*

I've even become a helping hand at the hostel, Angus has me running out to get the nightly video for all the tourists. The well-respected New Zealand movie '*Once Were Warriors*' has just come out, so each new tourist wants to see it, literally, every night. I end up watching it over and over before work.

It's the very light and fun movie about a man named Jake the Mouse who is stellar at molestation and wife-beatings... *Oh, yay.*

J.R. the skydiver and I have become quite flirty. He lives with Trev and Rien, and they invite me over for a barbecue. It's definitely a bachelor pad with a nice front porch, porno magazines, and booze everywhere. Trev insists on changing his name to Jake the Mouse. No more Trev. We get a Christmas tree for their home and they want to decorate it with our empty beer cans.

"That's funny, but I think we've all seen that so many times before," I tell them getting a bit tipsy myself. "What if we cut up your porn magazines and decorate with pictures of boobs?"

"My God, that's priceless!" Trev/Jake the Mouse shouts.

Everyone loves the idea. We get started with our scissors right away. The tree takes on the finished form with circles of nipples and tits, you know, *classy*. We decide the 'pearl necklace' picture is more festive for the topper. *Very wholesome for the holidays, indeed.*

It's almost Christmas Day and I feel like I know everyone in town. Even though it's summer in the southern hemisphere, all the store front windows have painted snowmen and snowflakes, Santa still in his red suit with the beard. I guess I'd thought they'd have a Surfing Santa, or a Bungee Santa.

I'm in my Taupo groove. At the market I say hello to Louie, my extra-tall boss, and then a quick hug to Michaela at the post office. I've met some cool locals—Clinton, Nigel, and Andrew who live down the street and come in every few nights for some beers. Another friend, Jasper, with the long curly red hair, is at the fish and chip shop and we stop for a chat. He's wearing a hoodie he made himself, beer-bar hand towels all sewn together. I can't look away. It's glorious.

I'm invited to a Christmas gift exchange at Mike's house. The skydivers, Victoria, Louie, and a few others are there. All the presents are gag gifts. Brothers Mike and Simon are notorious for consistently sticking fingers in their noses for any picture, even weddings, funerals, each other's noses, even friend's noses. So, they receive heaps of Kleenex boxes as their gag gifts. Next up, gloves with Velcro, and rain boots, *so while you're having sex with a sheep you have a good grip*. They even give me the gift of granny panties with a Holy Cow on the front. They actually fit and if you think I'm not going to wear these, you're crazy. I still only have my original pairs of panties from Africa.

On New Year's Eve the bar is beyond packed. It's so crowded, we just can't collect glasses or wash them fast enough. It's hard enough just squeezing through the crowd. At 10 minutes until midnight, just as a song is winding down to silence before the next song begins, Victoria gets so frustrated with everyone just barking their order at her, she slams down the tequila bottle on the top glass shelf. *Big mistake.*

In a fraction of a second, ALL the glass shelves shatter. ALL 50 bottles of alcohol plummet to the ground. Deafening glass fragments explode, splintering the mirrored wall into jagged shards, spilling tidal waves of pure booze down the drains at our feet.

We freeze, just stunned. Everyone blinks.

"Beer only!" Mike yells out, the music starts again, and we all keep moving like nothing happened. My God, that must be thousands of dollars just for the booze, let alone the broken mirror wall and shelves. Is she going to get fired? I guess not, no one speaks of it again. *So happy that wasn't my fault …*

One night at work, Sebastian's kind of cold and rude to me. I just figure he's in a bad mood from work when, all of a sudden, he *picks up a chick* right in front of me. They leave together, walking past me hand in hand. I sigh and shake my head. *Really?* I didn't think you were my boyfriend, but *really*, you can't at least meet her downstairs out of my sight? Glad we only made out a few times.

A regular friend at the bar, Tad, with thick curly gray hair and a Tom Selleck mustache, hugs me and says that's not very cool. Once I tell him we all share a room, he says he knows a friend with a room to rent. He brings me over to the guys I'd already met. Clinton with a mop of curly brown hair and bright blue eyes, Andrew, a big guy with the straight, dark hair down to his butt, and Nigel, my height with a buzz haircut and smiling brown eyes. All in their early 20's. They offer me the room tonight. I'll take it!

I get home 20 minutes before the boys to collect my things and check out of the hostel. Angus is there and he's drunk. He grabs my arm roughly and starts dragging me down the hallway, asking me if I need a proper bed for the night. I keep pulling back, saying no, but he's got my wrists now... literally yanking me toward what must be his own bedroom. *I think I'm going to be raped.* I plant my feet, elbow him as hard as I can in the ribs, yank my hands free from his sweaty, pudgy hands and run.

This night is terrible.

I have the address and grab a taxi the 15 blocks to my new digs. It's about 4:30 a.m. and I see the lights through the window as the new guys try to clean up and set up a bed for me.

It's a house with a proper flower garden in front, a trampoline in the backyard, and while the furniture is a little run down—unbelievably—three young men have furniture in every room. My room has faded wallpaper with roses all over it and two twin mattresses smooshed together with no sheets. And nothing else. I love it. Privacy once again. I set my bag down and unpack it right then, leaving folded piles on the floor where a dresser might go. I unpack my shampoo and toothpaste into my new bathroom. I grab a quick shower and fall asleep before my head hits the towel I'm using for a pillow.

"Knock, knock flatmate," Clinton says through the door in the morning. "Someone's here to see you..."

Really? Who even knows I've checked out? And even if it's someone like Sebastian, he wouldn't know where my new address is, right? Better not be Angus. I have no idea what time it is, but the sun is shining. I throw on a shirt and some shorts and go out into the living room.

SPLASH!!!

All three boys have full pots and pans of water and dump them over my head. I don't know the layout of the house and can't even see where to hide as they all keep getting me. I run out to the patio and Nigel turns the hose on me and it's *cold*. I'm squealing with laughter. When I grab a bucket and get into the groove of the water fight, they can see I mean business and soon, we're all drenched and in hysterics.

What a *fantastic* way to break the ice with a new person in the house. Finally, they make me waffles and we all get air dried lying on the trampoline in the sizzling summer sun.

I already like my new house. It's bright and sunny, the guys have rounded up a pillow and bedding for me. There's an old-fashioned wood burning stove in the living room. I settle in nicely, getting to know my new roomies. They roll a few joints and smoke, while I turn down the offer. It's Sunday and we're all goofballs. As they're cooking me dinner, we all get the same idea to smash into the small pantry, just to see if we can all fit inside. Yes, barely, if we all hold our breath... *cool, they're as weird as me*. The guys must work in the morning, Clinton's a crane operator and I'm not sure what Nigel or Andrew do yet, something in construction.

I walk the 15 blocks to work and it's not too far at all. That night after work it's a little scarier to walk that far alone, but I'll get used to it. As I'm ready to fall asleep, Sebastian comes knocking on my window. He's followed me home. He's completely drunk and sneaks into my bed and promptly passes out despite my disgust and protests. *Ugggghhh*. I grab my pillow and blanket and go out to the couch in the living room. I'd been looking forward to sleeping in for sure. I hope my roomies don't think this'll be a regular thing.

The next morning each of the guys gets coffee and breakfast at different hours and gives me weird looks, but they let me sleep. I'll explain to them later. Finally, mid-afternoon, Sebastian wakes up and has no idea where he is. He comes out into the living room and sees me, trying to apologize.

"Hey, Pam, so this is your new place, huh?" He says, almost like things went well on a date, and that's how he'd spent the night.

I don't say anything but turn away to make myself a cup of tea.

"Hey, about the other night, that's just traveling, you know?" he tries again.

Just traveling. You're either a douche or you're not, Sebastian.

"Sure... whatever," I say with a smile, opening the door. He gets the message and slinks out. He's already a blip in my rearview mirror, but every single time he sees that Bugs Bunny tattoo, he'll have to think of me *for the rest of his life.*

I jump into the shower. Halfway through, with shampoo in my hair, I hear a knock at the front door. Sebastian probably left his shirt or his keys or his soul behind. I yell, "Come in!"

A split second later there's a knock on the bathroom door as well. I poke my head out from behind the curtain and shout, "Just get what you need!" to which the door opens slightly, and a man pokes his head in. A strange man. With a ***POLICE*** hat on.

"Police, ma'am. Please take a moment and we'll meet you in the living room with some questions," he says and closes the door.

Holy shit! What have I done now? They must know I'm working with just a tourist visa. What are the laws in this country? I'm *busted.* And where is the pot in the house? Maybe they're into heavier drugs and I don't know it because I just moved in with three strangers. *I'm so stupid!* I'll get locked up abroad and never see my mom again.

I rinse my hair, quickly towel dry, and throw my T-shirt and shorts back on, hair dripping.

"Hello, officers, what can I do for you?" I ask tentatively, to the cops, a tall man and a short young woman.

They introduce themselves, explaining why they're here.

"We're investigating a stolen baseball cap from a neighbor who filed a report," says Officer Amy.

I'm wracking my brain. *Was there also a shoot-out? A stabbing? An assault?* I glance at their uniforms and see neither one carrying a gun. *Are they even real cops? Is this really an investigation? Are they really strippers and the new roomies are playing another joke on me?*

"I'm sorry... did you just say you're investigating a missing baseball cap? You're not joking?" I'm genuinely shocked. The crime is just so... *minuscule.*

"Right," says Officer Amy. "What can you tell us about Nigel, Clinton, and Andrew?"

"Well, not much I'm afraid," I'm truthful.

"Where do they work? Can you confirm their last names? How old they are?" Officer Thomas interrogates me.

"I literally just moved in here *yesterday*," I'm embarrassed, not for being questioned still dripping wet, but that I really don't know who I've chosen to live with. When I'm around other backpackers it feels normal, but to be around people with professional careers who live in this town, it sounds like I'm making it up that I've chosen to fly by the seat of my pants and move in with strange men. *Strange water ninja criminal thieves, apparently.*

"Ohhhh," they glance at each other, realizing I'm useless. "Do you mind if we search the house?"

"Sure, whatever you need to do... can I make you both a cup of tea?" I ask.

"Sure," they both agree.

They tell me the details of the *alleged* crime. It turns out a couple nights ago, the boys were at a house party down the street and Nigel was seen swiping a black baseball cap with the '*All Blacks*' Rugby League logo on it. Sounds hilarious to me. Why not just ask for it back, or maybe it was misplaced after a large house party, but whatever. Just so happy this isn't about me.

We get to chatting over the tea, and I break out some cookies as well. They ask me all about Hollywood, Disneyland, and I learn that no cops carry guns in all of New Zealand. They're locked in the car if needed. I ask what it's like *not* to have guns when you have to deal with bad guys. Especially as a woman cop. Officer Thomas states that Officer Amy's a real brave lady. He's seen her jump up on a 6'5" bad guy's shoulders and scratch his face, taking him down with a choke hold. Impressive. *I'd rather have a gun.*

The officers leave shortly after finding nothing in the house and tell me to truly enjoy my stay here in New Zealand. As I walk them out, they give warm smiles, hugs, and enthusiastic waves.

Not two minutes later, as I'm cleaning up the dishes, in walks Andrew, Clinton, and Nigel, who's wearing a black baseball cap with the '*All Blacks*' logo on it. They were watching everything from the neighbor's window.

I tell them how it all went down. When I say the crime was a stolen cap, Nigel picks up his hat as if to say, "Hello Ma'am" in a Clint Eastwood western film. "You mean *this* hat?" he smirks. We all roar with laughter.

I tell them I'm so used to American murders, Africa with its rapes and stabbings, Asia with the mass machete murder, this was *so very* hard not to burst out laughing at the very thought of a missing baseball hat.

"Good thing they didn't look in the stove," Clinton walks over to the wood-burning stove, long cold in the heat of the summer, opens the little door and shows all of us a bulging plastic bag inside. We can barely understand him between belly laughs, "Or else they would have found a *kilo of pot*!"

We all howl with laughter at the details. Me, caught in the shower. Me thinking they might be another prank. All of them spying on the cops searching the house. Nigel wearing said stolen property. And now the pot in the stove. Classic, and for reasons unknown, I think *these are my people.*

"And, Pam, did we just see you hug those coppers?" Andrew's still in hysterics.

"I did! What... no good?" I can't stop laughing.

*****

My flatmate Nigel takes me for a drive around the lake, showing me the farms one afternoon. We have to wait 20 minutes as a herd of sheep crosses the road. He tells me all about growing up in Taupo. It sounds like a lovely small town where all the people know each other. I tell him all about growing up in the suburbs of Los Angeles. He's fascinated that some of the Hollywood blockbusters were filmed in my home town, that we could go out dancing in Beverly Hills, and that we were about a 40-minute drive from either the beach, Hollywood, the snow, the desert, or the lakes.

The boys invite me to watch them play rugby. There's a deep, thunderous, masculine roar as they huddle together in the scrum. I can't even see the ball, ever. Having beers after, each of them with a bloody lip, or a black eye, I congratulate them on the game. "We lost," Clinton tells me. *Oh, oooops.*

I meet all their friends, including Mongo, a hulking blond guy who accidentally named his dog Grandfather in the Māori language. And Johnny, another blondie who had traveled to London. His funny story was that, while on the tube *(subway)* in London, he kept eying a girl because she looked exactly like Christie from school in Taupo. She kept looking at him as well, but neither said anything. About a year later, back home in New Zealand, he runs into Christie and tells her about being overseas and seeing her doppelganger. She laughs, "Oh, was that really you on the tube in London, then?" She was really there studying. Small world.

Today, we're all playing Paintball Wars. I'd played only one other time, six years earlier. I'd stayed alive longer than my friends and accidentally shot the owner in the face. So, I feel confident sneaking through the forest, sliding on my belly, alone for ages. I can finally see the empty fort with the flag on top. I wait, scoping out the woods, see no one, and make a run up the ladder into the fort. As I reach up to grab the flag, I'm elated! I'm a girl, and I'm going to win this boy-game!

BLAM! BLAM! BLAM! In that instant, each of the guys shoots my ass, I'm covered, bruised, and screeching out for them to stop. "I'm dead, I'm dead! You've got me! Knock it off!"

"We were laughing, waiting for the right moment to get you as you were sneaking along," Nigel laughs.

"Yep, we saw you literally the entire time, stomping through the bushes," Andrew headlocks me, laughing.

Clinton's in hysterics, "We were watching you think you were a badass!"

They take me swimming in the lake, to see the bubbling mud, to the Huka waterfalls, to hidden spots where only the locals know about, and mostly, they smoke a ton of marijuana. When they're high, the house rule is whoever stands up first *(always me)* gets shouted at: "Blonde with two legs!" and that means that person *(always me)* should make everyone a cup of tea with milk and two sugars. Occasionally they shout "Milo!" It's their Ovaltine-like hot chocolate milk... *party animals...*

The Holy Cow bosses have organized me to do "all the town has to offer" for free, take a picture of it, and then tell the tourists about the extreme sports, the sightseeing, and all the fun Taupo things. So far, I've been horseback riding in the forest, on a tourist boat ride, and today is bungee jump day.

I've always wanted to try bungee jumping. My friends and I have climbed up the long ladder and we're up on the minuscule platform, looking down the canyon over the turquoise river. Petrifying terror changes my mind. Toad, the main bungee boss, connects the straps to my friends' ankles, gives each friend a "Holy Cow!" sign, and the picture is taken. He counts down from five and, one after another, each friend just peacefully jumps over the edge, whooping with laughter. They're each gently lowered into the waiting boat.

Toad kneels to start connecting the cords to my ankles. I feel squeamish, even nauseated. My mouth is dry, my hands clammy... I feel faint.

"I think I'm going to skip it," I tell Toad. He glances up at me with his brown eyes, sun tanned face, his short brown hair tied in a rock star bandana.

"Oh?" he says, standing up, "Not your thing?"

"No," I smile, relieved that he gets me. "I'll take the picture though and still tell the tourists how awesome you guys are out here!"

"Cool, I was going to suggest that since you're all the way up here," he winks at me and smiles.

So, I waddle over to the edge of the tiny wooden platform, like a girl in shackles on death row. He hands me the sign and tells me I won't be so scared if I just don't look down. He says to put my back to the river, facing him. I smile, now calmly, for my perfectly posed picture. *Click.*

He gently begins to take the sign from my hands and in a slow-motion split-second, instead, yanks it from me with one hand. With the other, he reaches onto my chest and SHOVES ME.

Pure Panic.

I can't even wiggle my feet to try to find my footing. *What the hell? Am I even connected properly? Did he do a half-assed job since I literally just said I wasn't going?*

My face contorts into the moment of the greatest nightmare of my life.

My eyes bulge out of my head as I start rolling-down-the-windows with my arms, trying to grab at ANYTHING to make this *Not Happen.* I see Toad grinning widely, waving his hands around his face like a kid doing a *Neener! Neener! Gotcha!* He does a little dance as my head ducks upside down, below my feet, and I can no longer see the people on the platform.

The wind is rushing through my hair, my stomach is high up in my throat. I yell out, but not the normal high-pitched girl scream, like when I'm riding a roller coaster. This is a deep, guttural, man-moan yell from my diaphragm. I take another breath for another deep yell, and yes, another. GREAT, I'm still falling. The canyon walls blur. This is terrifying.

In the blink of an eye, I'm at the bottom. My eyes grab focus on the boat, when the momentum snaps my back like an old lady shaking out a bathmat. In slow motion again, I get whipped up a second time, almost as high as the first jump.

For just a hint of a second, I'm *Wile E. Coyote* after he runs off a cliff, his legs still running, despite the fact there's nothing more to run on. I fall again, with the same deep yell, another whiplash, and another bounce back up. And I'm spinning. Around and around at warp speed while dangling upside down. *Great.*

Eight bounces and continuous vomit-worthy spins later, I'm gently lowered into the boat with all my friends laughing and high-fiving each other. I'm traumatized. No high-fives. I cling to the edge of the boat for dear life. I could kiss the boat captain. Maybe I do, I can't remember anything but my one crystal-clear thought:

I'm so fucking glad that was free.

That night at work, my pictures are posted on the wall above the pool tables along with all the other shenanigans of the town. I guess this makes me a proper local. Everyone's congratulating me on my bravery. Little do they know I almost hurled the entire time. Toad walks in with the other bungee guys, runs up and bear hugs me. He explains I wouldn't have been up there at all if I hadn't been a teensy bit interested, so he'd really just helped nudge me along.

Yes, just like Mother Theresa, this guy. He's a real giver.

However, I do feel braver. I'm ballsier than usual tonight. I look across the crowded room, catching eyes with Simon, my boss's brother. He grins. I realize how handsome he is, instantly developing a crush. I've had it for, like, five minutes, but since I'm braver, tonight is my night. Simon's leaning down on his elbows over the bar with a bored look on his face. I walk straight up, don't say a word, and give him a sensuous kiss on the lips. His response? Blowing bubbles like a neighing horse. I pull back, smile and say in a sexy voice, "Wow... it's everything I thought it would be!" and walk away. The bored look never leaves his face, and his resting chin never leaves his elbows. No dice.

The night goes on and the bar gets to its usual full capacity, everyone dancing, people on tables. It's my turn to collect glasses, so I have about 20 mugs stacked high in my hands. I put a flimsy wine glass at the top as I head through the crowd to the kitchen. A tiny bump on my elbow from a nearby dancer rattles the wobbly wine glass. It shatters against the thick beer mug above everyone's heads as I pass through. To my right, a tall skinny guy quickly puts his hand to his face and there's blood coming through his fingers. '*The Talented Mr. Ripley*' shovel-to-the-face-*gushing*. A shard of the wine glass has sliced his face from his temple to his chin.

I shout, "I am *so sorry*! Are you okay? Can I get you some ice? A towel?"

The pulsating club music is too loud, and he can't even hear me as his friends rush him into the men's room. I find Louie and as I'm telling him to quickly check on my victim and his injury, we see his group of friends leave and hail a taxi... *to the hospital for stitches.* My eyes sting with tears. I can't believe I've scarred someone's face for the rest of his life. *I am mortified.*

Today is not my favorite day. At all.

*****

The next few weeks are pretty simple, how I see my boss at the market, run into Nigel at the bank, give hugs to Logi and Sebastian at the post office, Tad at the radio station, Michaela and Jasper at the Fish and Chip shop, cleaning the DJ's house. Man, this really *is* a small town. Each night is wonderful fun, dancing to the B-52's '*Love Shack*', now that I know the moves. Even helping to sell the bungee jumping, horse trekking, boat rides as I hang out some afternoons at the skydive drop zone and meet the tourists.

These divers have an unbelievable life. J.R. and Jake are 22 years old and Rien must be about 29, all from Holland. They each tell me that nothing is more exciting than going to work. There's a well-known locals-only joke that you should *not* be their first skydive because they'd just helped us close the bar a few hours earlier. They make over $100 a jump and usually get seven or eight a day. Must be nice.

One day, J.R. and the other divers take me to the De Brett's Hot Springs, a much-needed day with water slides and hot volcanic swimming pools. I feel like a high roller as we order lunch to our cabana. After the others head home, J.R. and I stop over at the deli counter when he notices a jar filled with imported Dutch black licorice for $10 each piece.

"How can I help you?" asks the chubby deli girl.

"The black licorice from Holland please," J.R. answers. "They're my favorite."

"Sure," she answers, getting her silver tongs out and pulling the lid off the large glass jar. Her hand hovers above the opening. "How many pieces would you like?"

"I would like all of them," he answers.

"Yeah! Why not, right? Me too! No, seriously, how many?" she laughs.

"I would seriously like them all," he smiles back.

"Sugar, there are a hundred pieces in here, at $10 each... that's $1000!" she replies, fully frowning.

"Yes, I would like them all. And the jar as well," he gets the money out of his pocket that he'd earned in less than a day and a half of having fun. She adds up the jar at $40 dollars. He hands the money over and her mouth drops open. I'm laughing. *I'm in the wrong business for sure.* We walk out in giggles, chomping on these wonderful European treats.

That night at the Cow, J.R. whirls me around the dance floor for a song before it gets too busy.

"Wow, Pam, you are the best dancer I have ever danced with in my entire life!" he tells me. I think back to my nightclub days in Beverly Hills and that it was important to me to know how to let the *man* lead. It apparently paid off. I get back to work and moments later, he shoves a note in my pocket. I pull it out and it says how I'm the best dancer with the date and his signature, so I won't forget. How sweet! It's too bad he gets together with a different girl every night, because there are some flirtations going on.

He comes back later with a video camera to film all his friends around Taupo. He's sending his parents a nice message and asks all of us to say hello. When he gets to me, I ask, "How do you say hello, you have a lovely son, in Dutch?"

"You say, 'Pijpen en doorslikken'," he tells me.

"Hi, I'm Pam from California, USA!" I smile and wave into the camera. "Pijpen en doorslikken!"

J.R., Rien, and Jake spit their drinks out. Jake yells out, howling with laughter, "You just told his Dad you want to suck his dick and swallow! Hahahahaha!" I have to laugh as well. I fall for everything.

The next afternoon at the skydive drop zone, I'm talking with Van, the owner of the field, and he tells me J.R.'s house had been raided by the cops late last night. All the boys were arrested for marijuana. *So happy I wasn't hanging out there.*

I go over to the picnic tables to sit with Jake the Mouse. We watch the plane take off, circle over the mountain ranges, with what looks like tiny ants falling, one after another, their chutes popping open. He tells me how it all went down.

They'd recently bought a kilo of marijuana for the three of them and any female guests who would partake in a toke. Late into the evening, there was a forceful knock at the door with the shout, "POLICE! Open the door!" *Hmmm, sounds familiar*. The cops searched the house with clear knowledge of copious amounts of marijuana being sold to these men.

Rien, J.R., and Jake were already high as a kite, so they were amused. They asked loud and clear, "Why don't you look in the closet? It's in the bin in the closet! The pot that you are searching for is up on the top shelf! In the closet!" But the cops kept looking all over the house, in the kitchen, under their beds... because yes, they were speaking loudly and clearly, but in their native *Dutch* language. These guys were laughing their asses off, literally telling the cops how to bust them in a timely manner. Now I *do* wish I'd been there, as a fly on the wall, to witness such hysterics.

Finally, an officer investigated the closet and pulled down the full bin. The cops insisted they must be drug traffickers as this is way too much for just the three men who live there. They were all arrested. Once they got their paperwork filled out, the boys had to sign their names, but instead, signed '*Horse's Dick*' and other obscenities, again in Dutch.

They went to court that morning and the judge angrily slammed down the gavel with a lesson like, "*Maybe you won't think this is so funny when you pay the steep fine of $500!*" which only made the boys laugh harder as they opened their wallets and had way more than that. J.R. even made sure to tell the judge he'd spent $1000 on licorice the day before.

The brightly colored parachutes guide them over, floating closer and closer to us, gently landing, with loud cheers, whoops, and hollers from the thrilled tourists. After everyone unharnesses, J.R. comes over, still wearing his

Superman costume, his daily uniform, and gives me a bear hug, spinning me around as I ask him what kind of bird doesn't fly.

"A jailbird!" I poke fun.

"No one ever speaks Dutch... it is way too easy!" He laughs with me. "I just wish I knew how to make it even more memorable!"

"Why not get the DJs involved and dedicate a song to the Taupo Police?" I suggest, still giggling.

"Of course! That's awesome!" He and Jake begin bouncing song ideas off each other. I mingle with the tourists and invite everyone to the bar later for some drinks, pool, dancing, and shenanigans.

We all stop into the FISH radio station and bombard DJ Paul, the absolute drop-dead sexiest man alive, *(*still using your words while helping with those memories, thanks Paul)* who loves the idea.

Over the airwaves, we hear *(*just told you: Paul's description)*, Paul's startlingly deep, manly, rugged, sexy voice "And this next song is dedicated from the local skydivers to the Taupo Police Department: Thank you for such an entertaining night last night! Making memories to last a lifetime!" while the reggae poppy sounds from the late 80's '*Pass the Dutchie*' blare out over the soundwaves.

I tell Clinton, Nigel, and Andrew about the adventures my friends had last night. They'd heard the radio dedication and already thought it was funny without knowing the juicy details. Does everyone in New Zealand have a personal *kilo* of pot? Big Mama from South Africa would do well here.

*****

From my kitchen window, I see the smoking volcano of Mt. Ruapahu, the snow-capped ski resort across the lake. It erupted yesterday, just lightly, I guess. It's summer now, so the resort part's closed but there's still snow and long, low, white clouds along the base. I'm domestically doing my flatmate's dishes since they'd just cooked me dinner. I'm watching the billowing gray ash and smoke, remembering my hike in Indonesia. It's beautiful, and it's causing the sunsets each night to be golden, purple, mixing reds with pinks.

My roomies take me to every rugby match, every barbecue, hiking, fishing, a drag racing thing, and teach me to drive a crane. They even take me hunting. I've never hunted before, but, when in Rome...

We head out with our rifles and I'm chit-chatting all day long about how beautiful it is, did everyone enjoy the sandwiches I'd made, and we even debate whether I'm telling the truth about the Circle K scene from the movie '*Bill and Ted's Excellent Adventure*' filmed two miles from my house in LA. The boys can't fathom the idea. When we end the day, I comment about not seeing *any* animals.

"That's because you wouldn't shut the hell up," Andrew says. "You scared them all away!"

Oh. Ooops. I didn't know hunting was a *quiet* thing. So, we try to shoot some fish in the river to see what would happen. Nothing happens.

I thought it was fun, so when Palmer, the skydive pilot, invites me for a midnight hunting session a few nights later, I'm ready for anything. Until he corners a shaking-with-fear-fuzzy-baby-bunny and pops a bullet into his brain, shattering his face off. I can't help it. I start crying and demand to go home right away. Palmer apologizes, saying he thought I'd like guns because I'm from Los Angeles and everyone has guns there...

Palmer explains that New Zealand doesn't have rabbits indigenously. English hunters brought them over 100 years earlier. They have no natural predators like hawks or snakes, and they're eating all the wrong vegetation. They humped like bunnies, and now they've taken over the country.

I don't care.

Dude, you blew up the Easter Bunny. I have tears running down my face as I admit I'd been relieved when we hadn't shot anything a few days earlier.

*****

The town of Taupo is getting smaller and smaller. Each day I seem to run into my boss Mike, his brother Simon, or at least five or six Holy Cow regulars. Swimming: Hi Simon. Grocery shopping: Hi Louie. Post Office: Hello Michaela. Catching the sunset at the lake: Hello Linnea and John from Sweden. Picking up a movie: Hello Mongo and Nigel. Over the last few days, I've run into my boss, Mike, about five times *a day*, all before spending 10 hours a night with him.

One morning I get a call from Winston, a boozy regular I barely know, to see if I'd like to go out on his boat with some of his friends to waterski. Yes, please! I'm excited to get out in a boat and I haven't water skied in a while. Plus, if I'm honest with myself, it would be nice to meet some *new* friends and have *new* conversations.

Winston picks me up and we drive over to the docks, waving to friends at every red light. As we park, he points out his speed boat and his friends are waving to us... his friends are my boss Mike and, yes, his brother Simon. I feel a mixture of unexpected hilarious disappointment, yet still a comfort. *My bosses are so cool they invite me everywhere fun. This is the 'Twilight Zone.' Our day will be filled with sarcasm and laughter as I'm sure to get a picture with their fingers in my nose at some point.*

Our day is spectacular, filled with laughter and even a tour of the lake. We head over to the west of the lake into Mine Bay and I see the impressive rock carvings, only accessible by boat. They were carved in the 70's by a man named Matahi Whakataka-Brightwell. He said the rocks called to him, so he carved out the Māori navigator Ngatoroirangi who guided the tribes to Taupo over 1000 years ago. Just stunning work. A Māori Mount Rushmore.

*****

I clean the DJ's house on the regular and get called in to do the voiceover work, so I'm a regular at the station as well. One day it's Mick and Jimmy's first day skydiving and it's being broadcast on the airwaves, with scorchingly hot and virile Paul (**you know why*) promising to bleep out any curse words. Well, there are plenty of curse words and absolutely no bleeping out. Everyone's too busy pissing themselves with laughter as Jimmy's squealing like a third-grade girl in pigtails all the way down. If it's anything like bungee jumping, *I can relate.* I vow never to sky dive.

Over the last few months, there's been a competition among all the groups pranking each other. It's all fun and games until Louie calls in a fake, anonymous *bomb threat* prank on the radio station, but gets caught and ends up in jail for the night. *Maybe time to take it down a notch.*

One day at the drop zone, I meet Scarlett, just after skydiving with J.R. She's a petite blond beauty with a little pointy Winona Ryder face, giddy and excited to hang out this evening. Scarlett's vacationing alone, from Australia, and it's her first time overseas. We bond. Of course she sleeps with J.R. and leaves on the Kiwi Experience tour bus to the South Island, just like they all do. But two days later, Scarlett has hitched back because she thinks she's in love with J.R. but, he now has Abigale from Russia staying at his house. *This is awkward.* Clearly, J.R. is not in love with her... or anyone. He's 22 years old, makes too much money, and is having a blast in a foreign country.

It's my day off today and I'm at the Crazy Horse bar having a beer with Tad, paid for by Sebastian. *This really is a small town.* Tad tells me he invented a part of a paint can with a circle rim inside, so paint won't drip out. Hmmm, random, interesting facts about my friends.

Scarlett comes in with tears in her eyes. She's been looking for me. She's just been told that no, she can't stay over with J.R. and he's very sorry, he had fun with her, but he's not ready for a committed relationship right now. She'd thought there was a spark and that after sleeping with him, he would ask her to move to New Zealand, and who knows, maybe they would get married and she would spend half the year in New Zealand and half in Australia, maybe move to Holland one day after the kids were born...

"Oh, honey," I say, giving her a big hug. "Honey, no. Oh, no, no. J.R. is wonderful, but he's not Relationship-Guy. You didn't just get dumped by your boyfriend. He's Travel-Fling-Guy. You don't cry over a guy like him, you enjoy your youth and adventure with him. You just banged a hot Dutch skydiver on your overseas trip. I'll let you in on a little secret... I've smooched him a time or two myself, and off he goes with a different girl each night." Tad nods, sipping his beer.

I continue, "See Sebastian the bartender? We dated a bit months ago, and he picked up a chick in front of me... and look, this town is so freakin' small that we have no choice but to be friends again." Sebastian hears his name, smiles, and winks. He probably thinks he should pick up Scarlett.

"Look, I was heading over to the Cow right now, but what if you come with me? I'll get you some dinner and we can have fun and drink all night and you can stay at my place tonight. I'm a bit of a vampire these days... it might be about four or five in the morning?" I tell her. "The skydivers will definitely be there though, so no drama for J.R., because he might have a new date...?"

"I promise," she says, smiling genuinely with big doe eyes and long lashes.

We walk over with Tad, who leaves early. J.R., Rien, and Jake show up with two Māori girls dressed in nothing but see-through pink lace nighties. I literally see their nipples as I say, "nice to meet you."

"They're hookers!" J.R. whispers in my ear. "I just picked them up at the liquor store and they have nowhere to stay tonight and no money. I told them to pick someone up at the Cow or stay with us at our place." He's genuinely trying to shelter them from having to stay on the streets.

He sees Scarlett and gives her a hug. He takes her out front to have a chat with her about No Hard Feelings. I think she "gets" him now, and to show him No Hard Feelings, she's flirting with him. She's even complimenting the hookers' night gowns. She must have had too many shots or something. After more drinking and dancing, about 3:00 a.m., we all leave the bar.

At J.R.'s place, the boys set up a mattress in the living room for the hookers to sleep on, while Rien and Jake smoke out with them. I can hear Scarlett calling my name from down the hall. I think I'm just going to walk the five blocks home, so I go to say goodbye. They're not in the bedroom, so I follow her voice calling me further down the hall, into the bathroom. Through the mist, I see the shower door is wide open, they're both standing naked, soaping up each other's bodies.

Do they need me to get them a towel? Do I owe them some money for the drinks?

A dripping wet Scarlett steps toward me onto the mat, takes my hand and gently lifts my shirt off, and in one fail swoop, J.R.'s behind me, taking my skirt down. They each pull me softly into their steamy, sensual world. I'm dreamily in the shower in the middle of their sudsy carriage wash. I'm just drunk enough to go with the flow, yet still sober enough to crack sarcastic jokes in my head.

Scarlett's kissing me and scrubbing my headlights. J.R.'s showing he has no hard feelings... oh wait, there is a hard feeling. She has a tiny body with huge, round, perfect tits... but what is that? Is she rocking a 70's bush? I need a weed whacker for your hair. How am I getting some action when there are literally hookers in pink nighties a few feet away? I know this is sexy, but can I really use some of that hair conditioner? It looks expensive. Wait, whose hand? I think I hear porno music in the background. Giggles. Laughter. I think Scarlett is my favorite girl's name. J.R. confesses he's never had a threesome before. Well, me neither! Scarlett either, so no one knows what we're doing. We're literally like the blind feeling the blind. Hilarious and awkward. Yet adventurous. Isn't everyone a little adventurous in college? Oh, yeah, I dropped out. Switch positions. Switch partners. Put your left hand in, put your left hand out, do the hokey pokey, and move your hand about.

We fall about with laughter, drying each other off. I run off in my towel into the living room and it's lights out, sleeping hookers, all tucked in. And skydivers each in his own bed, while Sugar-Joint fairies dance in their heads. I get us one large glass of water, and why not? We've shared almost everything else tonight. Scarlett makes sure there's room for me in J.R.'s bed, herself on one side and me on the other. We smooch a bit more, and slowly fade away into the night of What-the-Hell-Just-Happened.

I'm awakened with a throbbing headache, to the sound of scraping, squeaking, even... *grunting?*

I crack one dry, crunchy eye open and realize I'm in J.R.'s room, and oh, yeah, I had a three-quarters-of-a-three-way last night. We're all still nude in bed, all cuddled up and, oh, yeah, forgot to mention, there's SOMEONE CLIMBING THROUGH THE WINDOW.

I shake J.R. awake and pretend to still be sleeping. I hear J.R. talking with someone. Someone familiar, someone who is reminding him they have plans to go fishing this morning at 6:00 a.m., someone who can't believe what he's seeing, someone who sounds just like my boss Mike. Someone like Mike. Yes, My Boss Mike has climbed through the window and sees the orgy I just had. *Are you kidding me?*

This is a small town.

J.R. cancels the fishing trip and walks Mike out through the living room. Of course, he sees the hookers and naturally assumes that we all got in on some hard-core action. After Mike leaves, we all sleep through the afternoon, then grab some lunch out and laugh about who put what where, cracking jokes about our cracking jokes. Best three-way ever. I suppose, *only* three-way ever, but a perfect one, one for the books. *One for this book, sorry, Mom.*

We walk Scarlett to the bus stop so she can meet up with her tour bus on their way back to her flight home. She tells us this is the trip of a lifetime. We hug her and send her on her way. I head home for a shower. Yes, I need a shower after my *Shower.*

That night, before my shift, I get a call from J.R.

"Hey, Pam, I just spoke with Mike." He's very serious. "I told him not to mention what he saw this morning to anyone, especially not to you, as it would make you very uncomfortable. I told him if he even mentions one word of this then I won't bring another patron into the Holy Cow ever, nor will my skydive team. We will take each tourist over to the Crazy Horse and build their clientele. So, act like nothing has happened because I am not the type to kiss and tell."

Wow, chivalry is not dead. And this from a guy who signs 'Horse's Dick' on legal documents.

Now, I should probably have to move away and never show my face again, *however...* this could work out in my favor. I know my boss knows about something embarrassing and private in my recent present. Something I guarantee he's already told his brother, Simon, and his best friend, Louie. I could have fun messing with them.

I'm in my black shorts, my boots, and my Holy Cow t-shirt with my curly hair pulled back with a barrette, wearing lots of lipstick and mascara. I walk in smiling and saying hello to everyone. All eyes are on me, and one by one they ask, "How was your night off?" with a wonky look in their eyes.

"It was nice. I had fun, thanks," I say, grinning.

"Did you do anything special?" Louie asks.

"Nah," I say with a shrug, "Just, you know... *the usual...*" I turn away as all mouths drop to the floor. I bite my inner cheeks to keep from laughing. I think I've won this silent friendly battle until Mike begins the night as the DJ, and the first song he chooses is a new Jill Sobule song with the lyrics about kissing a girl... and she might do it again.

In sync, all eyes are immediately on me, then immediately averted.

Yes, everyone knows, yet can't say a word to me. Hilarious tiny little town.

*****

It's time for the sky divers to head back home to Holland. They decide to dress up in tuxedos and dive over Auckland for one last hurrah. Michaela and I are hanging out at their house while they get dressed. The boys are beyond high as a kite. J.R.'s explaining to me that anyone who is anyone knows how to wear a tuxedo; you have to button the inner button. And on and on he goes as Jake walks into the room, confused, holding up his jacket. I help with the precious inner button. Fits perfectly. Man, these guys clean up real nice.

Photos galore as we all tour the town, visiting the bungee jump site, stopping by the DJ's at FISH radio to say goodbye to Taupo on air, stopping by the Cow for a final table dance to "Love Shack," the Crazy Horse for a last shot, and finally the air field. These three will be missed.

At the drop zone, getting into the small plane one last time with Palmer the pilot, J.R. pulls me into a tight hug. He mentions if I want to call my family back in the USA, that the international long-distance phone line at their house will still be connected for another 10 days.

"Or call anywhere you want... we are leaving the country, and not paying the bill," he laughs.

"Thank you!" I shout over the engine. Michaela and I are hugging each other, feeling the end of an era. Linnea and John are tearing up, waving. Van and seven other divers say their goodbyes. We all wave frantically as the plane takes off. The door's still open, they're in their tuxes and parachutes, hanging onto the doorframe, climbing higher, all smiles, waves, and windblown hair.

They look so fucking cool.

Once they're gone, I head straight over to their old place and let myself in. There's still all the furniture, lots of coins left in the drawers, even food in the fridge. I kick back and dial my mom first, settling in for a long free conversation.

"Hello?" my dear mom's voice.

I have no idea what time it is over there. "Hey, it's Pam! I'm still in Taupo and my friends just told me to call anywhere in the world as they just left the country and are not paying their phone bill!"

"I'D BE VERY, VERY CAREFUL TALKING LIKE THAT IF I WAS YOU." *Says a very, very authoritative crystal-clear THIRD VOICE.*

"Pam, who is that?" my mom says.

"Did you just hear that?" I shout. "Oh, my God, Mom, I'm going to go!"

Big Brother monitoring International Phone calls. In 1996.

While it ruffles my feathers a bit, I go back for the next 10 days and call absolutely everyone I know. I certainly don't mention the bill, nor do I mention assassinating the President or kidnapping the Queen.

The town just isn't the same. But I still hang out at the drop zone in the afternoons because John and Linnea want to get their skydive licenses. Or sometimes I head over to the lake before the summer leads back into the fall. Nigel brings home a puppy that's only four days old. I make sure he didn't steal it from a party, then I make him send it back until it's at least six weeks old. I watch some of the rugby games. Some of us skinny dip after work in the lake and get out just in time to stop Simon from stealing our clothes. Luckily, there are empty Budweiser case boxes to hide behind as pictures are getting taken. The FISH radio station has the party of a lifetime at the station. My roommates and I even head out of town for some gambling at the casino for a weekend away in Auckland. Good times.

One night during a busy night at the Cow, a girl pulls me aside to tell me there's a glass stuck inside the toilet filled with urine in the ladies' room. *Yuck.* I ask her to pretend she didn't tell me, and I go about collecting and washing all the other empty glasses throughout the bar. Louie pulls me aside and tells me he must let me go. Yes, I'm being fired because I didn't reach into the toilet.

Well, this sucks. Too bad I wasn't a better worker. Like Victoria on New Year's Eve, and instead just smashed every single bottle of booze and shattered the glass wall. Then everything would be fine.

I head home and Clinton's still up and for the first time since I moved in, he rolls a joint and I smoke. We laugh for a while about me not reaching into the toilet. The next morning, Andrew gets up early and rolls me another joint and smokes with me for breakfast. Nigel comes home at lunch and he rolls me a joint and smokes me out.

"We all thought you were weird that you wouldn't smoke pot with us, not even once," he tells me.

"Yes, it does seem weird now that I'm getting high with all of you, all day," I tell him about quitting smoking pot at 16 and not doing it again until that night in Africa with the treehouse and all the animal eyes peering out of the bushes. We roar with laughter. He can't believe it.

"You should write a book," he tells me. *Hmmm, write a book? Me?*

They're home at separate times and roll a fresh joint. I smoke each time now. Why not. I just got fired. This goes on for *weeks*. Still night owls, one night at 2:00 a.m., there's a knock at the front door. All of us are watching a movie and shout out "Not It!" but I'm last, so I, therefore, am It. I open the door to find ruggedly handsome, chiseled, astoundingly sexy DJ Paul (**still helping with your description mate, we get it*) at the door asking for a coffee.

"Of course, come on in," I tell him. He'd met my flatmates before, so everyone gets energized at having someone famous in our home. While I'm brewing the coffee in the kitchen, Clinton hands him a Penthouse magazine to flip through. I return with the cups and he flips through the magazine while sipping. We all share a few laughs at the poses. He finishes his drink, says good night and leaves. We all look at each other and shrug. How does everyone know we stay up so late?

I haven't left the house in days. I'm just reading my books, lying out in my bikini on the warm trampoline in the yard, getting high three times a day. My mind's fuzzy. I keep re-reading the same pages. I'm getting paranoid.

Maybe it's time to head home to Los Angeles. What if my plane crashes into the ocean on the way home? Why am I traveling alone *anyway? I could get really hurt and no one would know how to reach me. Look what happened to my Dad and no one could reach me. I'm missing out on so much of my family and my friends lives back home. My ticket's no longer good from Malaysia. I'm trapped in a beautiful small town in New Zealand. I need to get another job. I shouldn't leave the house. Clinton will be home soon to smoke me out and Andrew will cook us dinner. Mmmm, food. I have the munchies. I'll just run out to the store to quickly get some chips and a Pepsi.*

I walk, leaving with $3 in my pocket, just wearing a tank top, some jeans rolled up and I'm barefooted. I run into Van, Linnea, and John at the market.

"Hey, Pam, where've you been lately?" Van asks. "Heard you got fired from the Cow. That's awful, mate. Have you found a new job yet?"

"I was just thinking maybe it's time to move on, maybe head home to LA," I tell him. "I'm just not sure what to do. I feel blue..."

He grabs my wrist and brings his face closer to mine until his forehead's touching mine, his long gray mullet tickling my neck. "Well, you haven't skydived yet. You can't go anywhere without your skydive!" He's pulling me with him. Our friends are shoving me towards his van, laughing. *Van's van.*

I sober up real quick.

"Um, I... um... I can't go. Now? Um, uh, uh... I... I don't... have any money with me!" I shrug like *Oh, man, otherwise I would totally go, darn it!*

"No charge for you, Pammy!" he tells me and pushes me into the car.

"I, um, I uh... uh, I don't... have any shoes—" I say quickly, in a panic.

"You can go barefoot, or use my sandals!" says Linnea, laughing.

We're driving away, heading over to the drop zone. I've been there a million times, yet always watching, never jumping.

"Wait! Wait! Just wait!" I shout, "I would need my camera for such an epic thing..." It falls flat as John hands me a camera, insisting I use his. I frown and take it. He grins.

We arrive at the zone and it's just past the last normal jump, so Van rallies everyone for a jump just for fun. There's a girl filming her boyfriend while he sky-boards, there are six others willing to go up, and even Palmer wants to fly once more. John and Linnea are going again too. There's a quick call made to ever so sexy Paul *(*yep)*, Jimmy, and Mick at the radio station and they're on their way over to interview me when I land.

As I'm getting into my space suit and harness, they give me a three-second tutorial about keeping my hands on my harness near my shoulders as we jump, and how to arch my back and let my arms out to the sides once we're free falling. *Or something. I don't know. It's all a blur.* I see a photo on the wall of Van and his four-year-old daughter, harnessed to her Daddy and having been on countless dives already in her life. I doubt that would be allowed in the USA, but hey, if she can do it, so can I.

We climb into the plane.

I sit in the co-pilot seat next to Palmer, who's laughing at the look I must have on my face. *Is this his revenge for my crying during hunting?* We take off, and no one grabs the microphone and gives that long, familiar speech about seatbelts. My flatmates don't even know I'm gone, let alone *skydiving*. Shouldn't I call my mom and tell her I love her? I look back and see all the other divers sitting on the floor of the plane. *Funny, I just assumed there were regular seats.* Along with goggles and space suits, we're all wearing foam helmets *because safety first, for sure.*

Someone slides open the door, and it's deafening as the wind pours in. Everyone just starts jumping out of the plane. I guess I thought there would be a pep talk, or each person counts to 10, looks back and gives a thumbs up. *Something? Anything?* There are 10 of us going, now there are seven, six, five, HEY SLOW DOWN or I'm next! *I still haven't properly decided yet...*

Van grabs my harness and gets me off my chair, so I'm on my knees across the airplane floor with him on his knees behind me. He hooks us together in a half-second and the last person goes out the door. They don't hover at all, just *whoosh,* gone.

My stomach is in my throat, my eyes as big as saucers, the wind whipping my hair like I'm riding a motorcycle at top speed. The clouds rush by, kind of like the top of that volcano I hiked in Java. Van taps my hands to cross over my chest and hold onto the harness. I sit on my butt at the open door of the perfectly good airplane with my legs dangling out. My reaction to sitting in the open door is to want to reach for the edges of the frame like a Garfield cat that doesn't want to take a bath. Hence the need to grab your own shoulders.

I think of the bag of chips and soda I was going to get. *I just wanted a Pepsi.*

Doesn't the punk band 'Suicidal Tendencies' have a song like that? Irony not lost while I'm leaning out of an airplane.

One more inch and we'll be out the door.

FUCK IT.

"Pam! Oh, no! Your hook! It's broken!" Van yells into my ear and we shove out the door.

We do an instant summersault and I see the plane fleetingly go from a full-sized plane to a micro-dot in the sky to show me just how fast we're falling.

My hook is broken? I know he's joking, and I can focus on the pit of my stomach that... somehow... *ISN'T falling.* WOW! This isn't the feeling I thought it would be... it's NOTHING LIKE BUNGEE JUMPING!

I DON'T FEEL AFRAID. I AM NOT FALLING AT ALL, IT'S LIKE I'M... I'M FLOATING!

Like the dreams I had when I was a kid that I was flying through the sky.

SQUEEEEEEEEEEEEEEEEEEEEEEEEEEEEEEEEEEEEEEE!!!!

THIS IS THE MOST EXCITING THING I'VE EVER DONE IN MY ENTIRE LIFE.

I'm breathing normally, not yelling or screaming at all. I'm squealing with *laughter.*

Van taps my shoulder and I open my arms, and now, I GET IT.

Life. Death. Love. Hate. Heaven. Hell. Earth. The Universe. String Theory. Time Travel. Quantum Mechanics. Quarks. 528Hz Healing Frequencies. E=mc2. Astral Projection. Interstellar Medium. Dark Chocolate. Puppy Eyes.

Zen.

I get all of it.

I feel the skin on my cheeks moving like a hound dog in the back of a speeding pickup truck. The wind is so much colder than I expected this high up. The ground still doesn't look any closer... it still looks like a beautiful green and gold farm quilt all sewn together. We're watching the sun set over Lake Taupo. Is that the area with the rock carvings from my water-ski day? And there's the billowing volcano still slowly erupting smoke, causing brilliant electric pink, purple, and red clouds off in the distance.

He points down and I see all our friends scattered within reach just below us. Van curves my arms to steer toward everyone. The guy from Israel reaches out to take my hand, we're flying even-steven. I reach out and we're holding hands. *Wooooo Hooooo!* Van forcefully yanks my hand back away from his and we re-set our course to watch the filming sky-boarding couple.

There are no words for the brilliance of the feeling, the majestic beauty of the landscape.

I come crashing back into the now, as Van pulls the ripcord and our parachute flies out and opens perfectly. My reaction is a gut wrenching *NOOOO! NOT YET! I NEED ONE MORE MINUTE OF THAT FREE FALL!* I simply cannot exist without more freefalling. I *must* do this again. Every single day.

We float gently for a while, and Van explains everything to me. Yes, the broken hook was a gag. No, I shouldn't have reached out to grab my friend's hand as we were traveling at 150 miles an hour and so was he, therefore at the point of our touch, it's a jarring 300 mile per hour crash site and can rip both of our arms off. *Lovely, you should probably mention that in the three seconds of training.* He recognizes the disappointment when the chute opens as well. Our free fall was much longer than for normal paying customers. They exit the plane at 9,000 feet and after a 30 second freefall, pull the cord at 5,000 feet. Because he owns the field and this was a fun dive, we went up to 18,000 feet and pulled at 2,500 feet. I got a 90 second freefall.

He even pulls the "brakes" and we're weightless for a moment. Zero gravity. It feels like we're training for outer space. God, I feel so lucky to meet the best people for the biggest adventures in my life. I'll never be able to top this day with these people. I do wish I had done it a few weeks earlier with the Dutchies as well, but today couldn't have been any better.

We float down and finally the ground begins to rush up at us. I kick up my feet and let Van land running on his feet. Then I place my legs down, trying to match his run. We stumble a bit as the parachute slowly drops in the wind behind us, pulling our bodies.

All my pals come rushing over giving high-fives and hugs. My cheeks hurt from grinning. I just jumped out of an airplane. I feel so high from life. Forget all the self-doubt I was having. I'm ready for new adventures again.

After unharnessing me, John hands me a glass of orange juice... skydiving sure is thirsty work.

"Wow, this *tastes like shit!*" I say. Linnea can't stop laughing and does a cartwheel.

I get home just in time to join the boys for dinner.

"Hey, Pammy," Clinton asks, "Where've you been?"

"Hey fellas," I am trying not to grin but can't stop myself. "I was kidnapped!"

"What the fuck does that mean?" Andrew asks, pulling his long hair into a ponytail.

I tell them how I was getting some snacks when my friends pulled up and forced me into the car and took me *SKYDIVING.*

"You mean you were at the drop zone again?" Nigel asks.

"No, I really went this time!" I squeal, shaking with excitement.

They congratulate me and ask me all about it. I keep turning the radio way up and they keep turning it way down. I'm rambling non-stop, extra loud. I can't sit down to eat, I'm not even hungry, I'm staring out the window, laughing, jumping up and down, reliving my freefall.

"Here, smoke this for Christ's sake," over my shoulder, Andrew shoves a massive joint into my face.

I can't stop laughing and they all seem a bit annoyed with me and tell me I'm spazzing out. Oh, yeah, they're just home from a long day at work and I've had the best day ever. *Sorry boys!*

"Come out to the Cow with me!" I'm practically yelling.

"No!" they each yell before going to bed.

I go to bed as well, and it takes me forever to fall asleep, but that's okay.

I know it's time to move on from Taupo. I need my adventures back.

The next day I head over to the Cow and everyone in town stops me along the street congratulating me and high fiving me. They've all heard my landing interview with the FISH Radio. I can't even remember the babble I must've said. Everyone at the Cow says how disappointed they were that I stayed home, that I am the first person *in all of skydiving history* to not be so excited that I needed to go out and celebrate.

I get an epiphany.

It's all that damn marijuana I've been smoking. It's like a vampire to my normal Me.

No more pot for Fox sake.

I tell them I'm thinking of leaving Taupo, but not sure where to go next.

Right then, one of the first guys I'd met in Taupo, Bennie from the Kiwi Experience bus, shows up and the others tell him what I'm going through. He offers me a spot *for free* on his bus leaving tomorrow morning for Queenstown, in the South Island. Of course, it's stopping at all the coolest tourist spots and most beautiful scenic drives, so it might take about 10 days to arrive. Would I like to go? Yes, please!

Having my last few beers with all my friends, I walk around town this evening and say my goodbyes. I genuinely feel the warmth of the people of this town, this home away from home. The universe just lined up and handed me a skydive and Bennie's offer to the South Island.

I tell the boys it's time for me to move on. They're all disappointed but they understand. We invite Johnny and Mongo over for a farewell hang-out.

"We're going to miss you," Clinton says. "Living with you was like having a crazy pet!"

South Island

Picton

- Magical Mystery Tour -
The Beatles

We take the beautiful 14-mile, three-hour ferry ride from Wellington through the Cook Strait. It's known for its choppy, rough crossings, making it one of the most dangerous passageways in the world. A few daredevils have swum the channel for 11 hours.

Landing in Picton, over near Abel Tasman National Park, we stop in Punakaiki to see the famous limestone landscape called Pancake Rocks on the edge of the sea. Hundreds of stacks of flat rocks, with waves crashing through blow holes and into surge pools. They're alternated in hard and soft layers of marine creatures and plant sediments under constant immense pressure. Did the ocean carve them? Maybe... *aliens*?

A group of drunk party animal Canadian girls keep asking me about living in Los Angeles and the celebrities I'd met in the past. I'd met many night club owners, who would introduce me to other club owners, who would insist I bring some of my friends in the next night. I was so poor I'd ride my bike to my lunchtime waitress job because I didn't have a car, but I'd ride in limos each night, getting into the clubs for free, VIP rooms, with free champagne all night. Then back to my hillbilly town.

Those were great times, I don't mind answering a bit about it, but they keep at it. I feel somewhat *cheesy* talking about it. In LA, you're a bit of a loser if you keep name-dropping. Plus, ZERO of these famous people would remember me.

I change the subject at first chance and ask them about their lives back home. Even that doesn't go over very well, because this girl has stories about her barn dance and that one has stories about her church bake-sale. They're embarrassed to tell me, thinking that I'm conceited now. For the rest of the trip we're still friendly, but now in a polite way, not in a really let's-get-to-know-each-other-way. I'm truly sorry it came up. We're not connecting.

And we are stuck on a bus together for 10 days. *Awkward.*

Franz Josef

- Wonderwall -
Oasis

We keep driving south through Westport and Greymouth. Eating, sleeping along the way, stopping to see the sights and grab pictures. Such verdant greenery out the windows of our lurching, lime green bus. I'm tired and it's cold by the time we pull into Franz Josef... our stay in what looks like a little Swiss town at the base of the Alps.

We can climb up the famous Franz Josef Glacier, it's *that* cold. Bennie had been very upfront that the ride was free but any excursions, hotels, and food, I'd pay for myself. I don't have as much money as I thought. I'm going to need a job sooner, rather than later. I pass on the glacier hike and stay inside by the fire, reading a book, drinking hot chocolate all day. Everyone on the bus probably thinks I'm weird, somewhat of a loner, since I don't join excursions or nightly drinking.

But if I'm honest with myself, that glacier climb would have been awesome. Everyone got to use pickaxes and wear crampons on their boots. There were hot pools involved. Did someone helicopter in just now? I'm a little grumpy to have passed on this one.

Seriously though, thank God that Andrew, Nigel, and Clinton each gave me a sweatshirt or a jacket. One's called a Swannie, basically a thick flannel pullover jacket. I'd brought tank tops and shorts with me. I was only going to Africa for three months ages ago. Now I'm expected to glacier walk?

Queenstown

- Plenty of Peaches -
The Presidents of the United States of America

What a *dream* of a town.

We pull into a hostel and, after checking in, I walk through the streets on my own. A *gorgeous* backdrop of blue sky, jutting mountain ranges with a rainbow forest. Every shade of green, with the mirage of reds, rusts, oranges, golden, and neon yellow leaves, surround the stunning royal blue and teal Lake Wakatipu. There are long, low, white cloud stripes nestled into the valleys and—wait—is that a clock tower? Why yes, it is! I turn toward the town center to walk along the Riverwalk. And is there an old-fashioned coal-powered steam ship riding along the river, like we're back in Mississippi in the 1800's? *Yes, there is!*

I walk along The Mall, an outdoor cobblestoned corridor with shops, bars, cafés and amenities like the pharmacy, post office, and such. It's alive with people. Some are tourists for sure, but most look to be locals and this is the town hub. I grab a Queenstown brochure and sit down at a coffee place called The Moa. In the guide, I see skydiving, *(Yay!)* Bungee jumping *(Boo!)*, speed boating, hang gliding, golfing, skiing, snowboarding, and much more.

I count my money. I gather all the coins and bills from each place I keep it, counting... 70, 75, 80... and—wait—that's it? Oh, man, that can't be right, can it? $80? $80! I need a job right away. Worst case, I can keep riding the Kiwi bus back up to Taupo and stay with the guys while I look for a job and back pay rent. I don't have a credit card. Okay, okay, I never got around to applying. I hadn't needed one before, I'd always saved and paid cash. *Oh crap.*

I glimpse my reflection after such a long bus ride. Hair wild and frizzy, not a stitch of makeup on, a boy's shapeless Swannie that's a faded cat-barf brown and grey flannel, jeans with duct tape on the ass and knee, with worn out men's black steel-toed boots. I haven't plucked my eyebrows in weeks. I am a hot mess. *Sigh.* I walk into the first bar I see and ask if they need any help as a waitress or bartender. The manager and bartender look me up and down and tell me to try Chico's farther in The Mall. Ask for "Spud" and "Skanker" I'm told.

"For real? *Spud and Skanker*? Those are their names? Maybe you're making fun of the tourist?" I ask with a grin.

Nope, they're the owner and manager, and one of the waitresses just quit.

I walk over to Chico's and ask confidently for "Spud" as I simply refuse to say "Skanker." A tall guy in his mid-30's with light brown hair and big round blue eyes comes over. He gives me the onceover. Can I start tomorrow? I cannot believe my luck. And I'm a *troll doll.* He gives me a uniform T-shirt and I start walking back to the hostel. The mountain range is turning a bright bubble gum pink in the sunset. It's absolutely remarkable.

"Hey, is that you, Pam?" I turn and see a tall, skinny, sandy-blond mullet heading my way, attached to a face I've *never seen before in my life.*

"Yeah, hey, umm...?" I smile.

"Jason, from Taupo. I was a regular at the Cow? I was the one who did the bungee run for the beer that time? And you would dance on the tables when the *'Love Shack'* song would come on?" *Well, at least he knows enough to be telling the truth.*

"Oh yeah!" I fib, no idea who he is. "What are you doing down here?"

"I moved a couple of months ago, how about you?" he asks.

"I literally just arrived an hour ago, I just got a job at Chico's!" I tell him.

"Where are you staying?" he asks.

"Just around the corner at the hostel," I reply.

"Hey if you wanted to save some cash, you could crash over at my apartment?" he tells me, hands up, like he's carrying an imaginary pizza.

Hmmmm... save money, stay with a strange man... save money, man, money, man, money...

"You're sure? Maybe a night or two while I find a place?" I cave in.

"No trouble at all!" he explains. "In a few days, I'm moving to another place just on the outskirts of town, so my place hardly has anything in it, but there's a mattress on the floor in your own room, okay?"

Well, he seems nice enough. We walk back together, and I grab my backpack. I check out of the hostel, getting my money back of course, which brings me back up to $105, and I'm strangely relieved to be back in triple digits, as if I'm not clearly still in poverty.

"That's all you've got?" Jason asks, puzzled at my tiny backpack.

"Yes, not bad for a girl, huh?" we laugh.

A few blocks later, he opens the door upstairs at his place. It's a pretty small one bedroom, a crazy Las Vegas turquoise carpet with streamers and confetti on it. He'll stay on the couch where he says he sleeps anyway. The mattress is bare on the floor in my room, and he gets me a pillow and blanket. This will be great!

I sleep most of the day and take pride in getting ready for work at Chico's Bar and Grill. Long hot shower, extra conditioner, and man, I notice my hair is down to my butt when it's wet. I've been trying to grow it long for years. I pluck my Frida Kahlo brows and... okay, okay... I might even hypothetically Shave my Mustache. *Allegedly*. Lots of mascara, eyeliner, blush and lipstick, curly hair like Julia Roberts, the top pulled back loosely in a golden barrette. I pull on my plum-colored T-shirt with long black pants that I recently found for a dollar at the Goodwill shop.

I walk to work through this amazing town. It takes my breath away again, the sunset turning the mountain summit a bright pink salmon. I arrive right on time. There are two big roaring fires in the large stone fireplaces on either side of the restaurant. Yesterday, Spud had said he begins everyone's shift with a free meal, so I'm showing up for my breakfast.

Spud doesn't recognize me. He frowns with confusion. He can clearly see I'm wearing the uniform shirt, yet I have to remind him he'd hired me last night. His face lights up with approval as he sees the Pampered Pam.

"Any better?" I ask him, laughing.

"Just a bit," he says with a wink and a smile.

"Hey, I just got off a two-week bus ride!" I pretend to be offended.

He asks all about it and is happy to hear that I was a local in Taupo, he's got lots of friends there. Spud introduces me to some of my new work buddies. Big Jude is an enormous Māori bouncer and Joff is almost too young to be the manager, with his blue eyes, rounded cheeks, and blond curls. Quiet Seth is the bar DJ until the professional DJ arrives, and Big Ted is the tall, green eyed, gorgeous bodybuilder. And Skyler, she's from England and traveling as well. Skyler is a petite brunette with sparkly, giggly brown eyes and a smile that invites you in for a cup of tea. I like her right away. Then there's Razza. He's the Canadian Dane Cook look-alike, and he's quite the comedian. In the kitchen is Sawyer Dog, a sarcastic surfer type.

We all sit down to eat like a family and it's... liver and onions. I dig in, not to be impolite, but this is my breakfast... *eeeewww*. Everyone makes me feel right at home. It turns out Joff and Skyler are dating, and he even knows Simon from the Holy Cow. Big Ted makes small talk with me. He says he was working last night when I came in and applied for a job in a Swannie. He laughs and laughs.

"I just could not believe it! A girl... applying for a job... in a SA-WANNIE!" he bursts out laughing with his mouth full of food, spraying the table. I laugh along, but seriously have no idea what I must have done that's hilarious.

"That's a first in the history of New Zealand for sure, Pam," Spud chuckles.

"My flatmates gave it to me when I told them I was heading to the South Island," I giggle along, mentally stabbing Clinton. "So, it's only for, what, like, running to the market or something?"

"No, it's only for working on a farm, shoveling horse manure... in the 70's!" ribs Razza.

"No—seriously, Pam, you clean up real nice," Big Ted tells me.

I smile. Everyone wants to know about America, Hollywood, Las Vegas, Disneyland. After we're done, we clean up and the crowd begins to walk in and walk in... more people. I start straight away, taking my first order like I've been there a year. In the USA, I'd have had to train with another waitress for a week. This feels better. Skyler shows me around and trusts me to get the job done. Later, we clear out the tables and the other DJ arrives to get the crowd on their feet. This place has quickly turned into a pumping night club. Skyler and I wear little packs to take cash, and hold drink trays high overhead, collecting glasses. The music's loud and it's like the Holy Cow's energy, so I'm in my element.

We finally stop and take a breath at 5:00 a.m. We're *starving*. We last ate 11 hours ago. After cleaning up, the gang shows me the first shop open is a bakery, and they sell last night's meat pies for a dollar. I dig into my beef hot pocket, we share some laughs, and head home. I sneak into my apartment and am so tired I sleep until late afternoon. As I shower up, the heavy smell of cigarettes seeps out of my hair, out of my skin. And I'm back at Chico's at 6:00 p.m. for a breakfast of kidney and potatoes. *Yum... no, not really.*

This goes on for a few days until Jason's moving day. He mentions his new place might also have an extra room for me. It's a good four miles out on the edge of town though, but I say yes for now. I pick up my backpack and we drive over the river, and through the woods, to scroungy, stinky, loser's house we go. We arrive to a run-down barn and an even more run-down house with a couple of old red-neck men. They lead me to a clean enough room, with a bed and some blankets, and a lock. I'm friendly but not loving this. To be honest, I'm not really feeling comfortable, but I leave my backpack to catch a ride to work with Jason.

Another great fun night meeting all the people in town and sharing laughs with my new crew. I was planning on taking a taxi back to my new place, but it turns out Big Ted lives all the way out near there as well, I can catch a ride with him. In a canoe. At 6:00 a.m. Yes, he's been canoeing to work and back along the breathtaking river with those mountains in the view. I'm chomping away at another meat pie, this time with melted American cheese. Ted's eaten about three tuna sandwiches throughout the night.

"Gotta keep up my protein!" he tells me, spitting tuna on my cheek. "I don't know how you all can go all night without eating. It's terrible for you, so unhealthy!" *Mmmm, meat pie.*

"The mountain range is actually called 'The Remarkables.' They change colors all year," he's huffing and puffing with the oars, gliding us smoothly across the glassy water. The sun's rising, but it's freezing out. I can see my breath and my nose is Rudolph red. The leaves are falling all around us, dancing on the water. I'm half expecting a crocodile to pop up out of the water at any moment. Then I remember I'm in the safest country ever, where nothing's venomous or dangerous. All I need is a parasol and Ted to serenade me with a ukulele. I love it here.

I thank him for the ride and walk the five blocks home. As soon as I'm inside, I'm exhausted but I notice *these men have been through my backpack*. My room still looks the same, and nothing's missing, it's just all shoved back in, all muddled, not my style. I'm too tired to care. But will completely keep looking for a room in town. I shower up and hit the sack. I should tell Jason when I see him next. We're on opposite schedules as he's out of the house before 6:00 a.m. and I leave for work right as he's off.

The next night, I'm speaking with a regular customer—Denny, a tall, skinny, dark-haired guy, and he can't believe the guys have gone through my bag. He lives with an older lady and calls her to see if he and I can share his room with twin beds, until I find somewhere safe. I jump at the opportunity, grab my bag that morning, and head over to Denny's. His flatmate is lovely, offering me tea and chit-chatting about America. She even has a hairdryer I can borrow while I stay. I've had curly, frizzy camping hair for over a year on my travels, but for 15 years at home I straightened it. Plus, all my travels have been in warmer climates, but to leave the house with cold, wet, shower hair in the freezing temperatures is something I might have to rethink.

I head over to Chico's with my sleek, smooth, and very long hair. Everyone compliments me on my new look. I feel the most like my old self... it's been a while. A while since I cared about my looks, which has been quite freeing. A while since I've felt like the old me, which, mixed with the new wiser me, feels a little bit like I'm a double agent spy for the KGB. *If I'm a cool hippie chick now, is it treason to still be vain? There's no hippie-how-to-handbook.*

Denny is friends with Emmit, the dreadlocked, missing teeth, fun drunk. He points out the much older, *only homeless guy in all of Queenstown*, also with dreadlocks and missing teeth, who comes in every night and steals the drinks people leave behind while they dance. Everyone in town knows this and doesn't care. They let him in anyway just to get out of the cold. There's also the Wayne Gretsky lookalike, and Tama the Muscle Māori local with the biggest, warmest smile.

And I finally meet the infamous Skanker. He's much shorter than I am and has recently quit the bar business, but he's still welcome to all the perks. He gets so supremely wasted that his head's bent over the bar, cradled in his arms. It looks like he's passed out standing up. He pukes all over his shoes, looks up at me and says, "Three more whiskeys!" I notice his T-shirt just says "Bent" on it. I tell the other bartenders what just happened and how he's re-ordered. No one even blinks an eye as they serve him the three more drinks, even while he's still standing in his own vomit.

I've been looking in the paper for a room to rent, so Denny takes me the seven blocks up the hill overlooking the lake to check out the room at Chad's place. Chad seems lovely, a handsome guy with the usual Kiwi look—brown hair, green eyes, a smile, and a flannel shirt. He greets us both as though we're a couple moving in. He works at Skipper's Canyon, strapping in the bungee jumpers, and offers us a jump anytime.

"No thanks!" I yell out too fast as I explain my story of being bungee-pushed. We share a laugh as he shows us around. It's a cute two-bedroom cottage with a small porch and foresty backyard. Inside there's a warm wood-burning metal stove in the living room, and as soon as I see the bedroom with its peach-colored walls and burgundy bedspread, with a view of those "Remarkables," I know it's going to be a fit. We shake on it and I gather my things from Denny's place. He was very kind to share his room with me.

I call home and my mom is so pleased to hear that I have a new job, a new home, and how much I'm loving New Zealand. She puts my sister Sandy on and, giggling, she tells me she's buying a ticket to visit me in Queenstown. We chat for ages. I'm giddy with excitement that my sister's willing and able to meet me halfway around the world. She can stay for two months, so we decide a month in New Zealand, then back to Australia for the second month. She'll arrive in about three months. She'll even bring my snowboard.

Some new friends want to show me around town, and of course I agree. *Thanks, Rod Stewart.* Finn takes me out on his steamship (Yes, that steamship), on a cruise around the lake. The *TSS Earnslaw* was built in 1912 and ferried supplies to the needy back in the day. He shows me around the ship, we have lunch. Down in the engine room, I help him do his job for the day, officially, I'm a coal shoveller, feeling like Mark Twain himself.

Next stop is Walter Peak Farm where they shear some sheep for us tourists. The farmer walks up, leading a large bull, and I hear the question "Who wants to volunteer to ride this bull?" My hand shoots up before I have a chance to think about it. I put a foot in the farmers clasped hands and, yep, I'm on top of a real live bull. I'm a bull rider. *Thanks, Rod Stewart?* The steed stomps around while the farmer tells us all about him. I can feel him breathing. The bull, not the farmer. A memorable Queenstown day.

I meet another group, a gang from London... there's Noah, with his black wavy hipster hair to his shoulders, and his blue eyes, he looks like a model. His best friend is Panda, with his sandy shaggy hair and his heartwarming smile. There's tall Rick who thinks he's the bee's knees and who's been all around the Middle East. They all share a house down the street and have found jobs in town. They've met while traveling too. They come in every night for a few drinks and we share our stories.

I'm quickly falling into a routine of vampire nights and days, eating mutton and meat pies with 11 hours in between. Can I get a piece of fruit anywhere?

Seriously, if I could just eat an apple, or just one peach. Even a bite. Haven't even seen a peach in a year.

I sit at The Moa, with my chai tea latte each day before work. Listening to the saxophone player busking for tips along the cobblestone promenade, while the crisp air turns downright biting. Winter's upon us now. My bedroom has turned into an icicle each night, so I try to help Chad chop wood. I quickly discover that's a boy's job for sure. I agree to clean the house properly if he can keep that stove lit. I want to buy an electric blanket, usually $10 in the USA, but all I can find is a $70 heating pad with three settings: not frozen, lukewarm, and room-temperature. It'll do.

Razza's little brother Tyson arrives from Canada. He immediately becomes a center of entertainment, with pictures of him hanging off the roof, or naked in the kitchen with his nut sack shoved into a mug. We nickname him Tazza.

One night, I meet a couple of Australian tourists from Sydney. Brett and Jerry, early 40's, both wearing glasses. They arrive for dinner early so I spend some time with them, talking about things to see and do here. The next day's my day off and they invite me for a hike in the afternoon.

We meet at The Moa and get started on our hike after a hot cocoa. Coming from Los Angeles, I ask the typical first question. No, not "what kind of car do you drive," the other one, "So what do you do for a living?"

"We're both geologists," Brett tells me.

"Oh, how... um... *boring*!" I say with a laugh to show I'm just ribbing them, but I really don't know what a geologist does.

"Oh no," Jerry says. "I can tell you everything about this planet. From studying the minerals in my wedding ring, I can pinpoint the exact mountain where the gold had been mined, anywhere in the world."

Brett's also excited, "And look here," he says pointing to the lonely black rock about the size of a football. "This is a volcanic rock that should be from an ancient Taupo volcanic eruption or even, if I studied it further, maybe it was from that famous eruption in Indonesia ages ago, called Kra-"

"Krakatoa?" I jump in, excited now as well. "I just traveled through Indonesia! I ferried right through where Krakatoa used to be. I just learned all about it last year."

"See?" Jerry says, "We love it. It explains so much more about the world than people ever realize..."

"Did you know the snow we have now isn't even supposed to be here?" Brett tells me.

"You mean, like... in Queenstown?" I ask.

"No, I mean the entire ice caps, the North and South poles," Brett goes on to explain. "From things like ice-core dating we can read the carbons and the history through the layers of time. This *entire planet* was 95% swamp and rainforest. From whatever caused the ice age, either a meteor striking the Earth, or if our planet wobbles off its axis once every so many millennia, or a flare from the sun... the poles are melting because *this planet was never supposed to have snow in the first place.* It's slowly going to disappear. You'll see. Didn't you learn about greenhouse gases in high school science class?"

"Yes, sure, but no one told me all this!" I'm fascinated.

"And things like our cars, hairsprays, and even McDonald's are helping to speed it up faster," Jerry says.

We stop along our path overlooking the lake, the Remarkables. There's the little steamship putting along below us, with little puffs of smoke coming out the pipes with each shovel of Finn's coal. We sip our waters as we rest a moment.

"McDonald's? How so?" I ask, happy to have my own private tutors today.

"Well, McDonald's is one of the highest offenders because they mow down entire forests to make room for the cows to graze for the burgers, plus every cow puts out quite a bit of methane, to say the least," Brett looks saddened as he goes on. "Plus, as human overpopulation occurs, it'll keep going exponentially faster."

Jerry jumps in, "You just went through Indonesia, right? You see how everyone burns some trash, but also leave piles of trash? They also burn the peat moss from the forest floor, which holds thousands of years of CO_2, so just one little intentional forest fire can release a millennium of toxic gases. First world countries are even worse. We know what we're doing to the planet, but we just don't care. If third world countries don't educate their people about the world, and big Western corporations don't stop, we're all going to be in big trouble."

I'm stunned. I knew some of this, but not the details of how it all fit together. I'm sad for my future children's burdens. I hope my generation figures this out before it's too late. I'm grateful to have met them both.

One night at Chico's, a customer with a familiar face walks in. Simon from the Holy Cow, along with some new friends. They've all been riding the Kiwi bus. I turn to introduce Joff, but Simon sticks his finger in Joff's nose. I'd forgotten they're old pals.

They introduce me to everyone. Heather's one of the new friends, and she's already nicknamed 'Heater.' She's English, with chestnut curly hair, bright blue eyes, a cute freckly nose, and pierced lip. She's going to stay after the bus leaves and make this her home base for a while. We instantly bond, sharing a laugh with Simon as he pretends to give his friend a blowjob, smearing mayonnaise all over his face for the happy ending. Appropriately, Simon's shirt says "*You're staring again!*" A few days of fun around town with them until Simon heads back up north.

Heater rents a little camper van in some old lady's driveway. She and I become best buds, sharing our life stories, and hanging out at The Moa each day. Saxophone music wafts through the air, and we're sipping our warm teas, just people-watching. From her village in England, she'd bought a ticket halfway around the world to ski in New Zealand. She's 22, wears purple velvet bellbottoms and can roll a joint with one hand... this is one cool chick.

It gets colder and colder as the snow begins to fall, causing a light dusting over the beauty that is Queenstown. Heater tells me each morning the glass of water next to her bed has frozen solid. Every night she's at my bar, mingling with everyone. The weeks go by and we hang out with Noah, Panda, and Rick the most, with dinners at their house along with a hilarious card game where whoever loses is forced to do the house chores. No excuses. I've had to vacuum, while Heater must sew a few buttons on some clothing.

*****

The music. Quiet Seth does a great job playing CDs until the DJ arrives who then plays the exact same songs each night, over and over. First, "Stuck in the Middle with You," "Plenty of Peaches" "Timebomb," then Grease—"You're the One that I Want." I hear the same songs at least five times *each night*. In the same order. There are long silent breaks in the music while he changes the disks. It's not good. I ask about it.

"Keeps everyone dancing though!" he's proud of this. No dude, they want to dance. They're having fun *in spite* of you.

"Okay," Seth tells me, after hearing my ninth complaint. "You're the bar DJ tonight, until the other DJ arrives. Let's see how well you do." *Yikes, what have I gotten myself into? I'm heavy metal girl.* I know dance songs when I hear them, but I don't know who sings them or what's the band's name.

Ummm, there's Ziggy Marley. I throw on "Tomorrow People," and everyone dances. Okay, I can do this. The song must last, what? Like, three minutes? I've got to search for the next one, and seamlessly switch. I can barely see the equipment or read the CDs in the dark bar light. I pick out "What is Love," "Rhythm is a Dancer," "Hotstepper," and "Boombastic," even Beastie Boys. Hey, people are dancing... but I'm out of ideas. I find a few more songs, but now I just freeze. I beg Seth to take over. He laughs and does so, seeing my panic. He does the songs while also playing bartender. I could barely cover just the music. I'll admit, it's harder than it looks. I'll stop complaining.

Winterfest is coming. The entire town will compete in funny, wintery games. They'll wear costumes, do races, and swim in the lake that only changes temperature by two degrees, even in the summer. It starts off with the scantily clad "Top Bloke" contest tonight.

And just in time for the fun, my sister Sandy is arriving!

Heater and I are off to the airport to pick her up. She'd traveled through Europe for a few weeks with our high school pal, Mark, while I was in Africa with Jay. We both thought I'd be gone for just three months. I'm impatient, I haven't seen her in over a year and a half. The longest we'd ever been apart.

In the pickup area, I'm actually nervous. I see Heater looking past me with a coy smile, nodding. I spin around and there she is.

Sandy! She's got long, sand-colored hair, green sparkly eyes, a smattering of freckles across her nose and a little, super-fit Jennifer Aniston body. She runs toward me. I lift her off the ground and spin her around and she does the same to me. We hug, we cry, we laugh. She looks exhausted after her long flight.

"I think I saw a billboard of you bungee jumping!" The first thing out of her mouth.

"Wait, what?" I ask in disbelief.

She had to change planes in Auckland and there was a billboard of a girl that looked exactly like me bungee jumping in Taupo, she even took a picture of it. I can't say for sure because there's a glare from the lights right across the eyes, but that would be *comical.*

We're taxiing home, and she's exhausted. But, well, here's the thing... it *is* the "Top Bloke Contest" tonight and she wouldn't want to miss *that.*

Poor thing, we let her set her bags down in my room, meet Chad, shower up, and drag her out to meet everyone.

The entire town is there. As we sit down to a large table at the event hall, everyone's helping themselves to the buffet. There's a catwalk and Razza is the emcee with a microphone, with Tama, big Ted and 25 dudes all dressed only in their underwear, each with a bowtie around his neck. It's hilarious as the men do a talent show, flex their muscles, and crack up the town with laughter, all while trying to out-humiliate each other. As the winner's about to be chosen, our Chico's table gets loud, chanting "Tama! Tama! Tama-Tama!"

We even get Sandy to chant with us, drumming on the table to match the beat. It works. Tama's crowned the "Top Bloke" of the Year, Mr. Winterfest of Queenstown. I stand up to clap, while Razza's little brother Tazza jumps onto my shoulders, like he's a four-year-old and I'm his Mommy. Somehow, my knees don't buckle and I'm holding up a full-grown drunken Canadian man on my shoulders. A terrific way to introduce my sister to the town.

Sandy and I sleep in late, cuddling for warmth in my bed. The first thing we do when we wake up is unpack. Sandy brought my snowboard, a pile of jeans to sell to friends. Hey, they're a few bucks at the vintage shop in LA, but can sell for $90 here in New Zealand.

We talk about Dad's funeral. She tells me that, when our mom went to his apartment to gather his things, she found bottles of prescription pills everywhere. It turns out, the crazy behavior leading up to his passing was due to his hidden pill addiction. One single doctor approved all his fake pharmacy transactions. I'm torn between my father's personal responsibility, and the pure hatred of his doctor.

Someone my Dad trusted. Someone who knew what he was doing. Someone who killed my Dad. Tragic.

We talk about the good old days as a family. It's the first time I can truly begin to heal, with someone who knows me, knows our family.

She then hands me my most treasured gift of all time. She hands me a photo album. She's taken our entire lives and "Creative-Memoried" it. The kid jokes, there we are at eight years old in our jammies, singing into the vacuum cleaner handle. There we are swimming in our pool, partying as teenagers, dressed up on Halloween, on our first trip to Mexico without parents.

I'd had my entire life stolen from me just two years prior. The only thing I'd have grabbed in a fire would've been my photos, yet that's the only thing the thieves threw away. I'm truly touched at the thoughtfulness of this gift of my life in a montage. My eyes well with emotion at the thought that my sister has lived this life with me, and she is now sharing her photos, her memories, that are mine as well. We decide to go make more.

Winterfest is all week, so we head down to The Mall for the fun. The band's music is pumping loudly by the lake, and we can hear it for blocks as we walk. We find Noah and Panda and their gang all dressed up as cheeky pirates. The first contest is with BMX bikes, tricks, jumps, and flips off the dock, landing in the lake. Next, there seems to be a couch race down the snow, where everyone in costume pushes their old couches down the groomed street, hopping on to cross the finish line. It looks like Gretsky and Sawyer Dog are in that race. An archery competition, a water balloon toss, with music and bonfires in the town square, Sandy has chosen the perfect time to enter my life Down Under.

I take her to The Moa for hot coca and the saxophone guy's there playing his little heart out. She fills me in on the details of her trip around Europe last year with Mark. We all grew up together. He'd taught both of us to snowboard, kayak, rock climb, and we shared countless adventures and camping river trips over the years. Everyone in town knew him, so literally everywhere we went, we heard "Hey Mark!" and—yep—it was a friend. But he's CRAZY. This guy graduated high school in a wheelchair after jumping off a roof into a swimming pool. He'd missed completely and broken his back. He was told he'd never walk again. Not only did he walk again in a few months, he taught us all those sports *after* the accident.

They'd decided to go on their trip after I was already in Africa, so they had to learn how to travel together too. They too didn't know how to leave the airport, and, after watching them wander around for hours, a lady had to help them. *See... that would have been me.*

Mark ran with the bulls in Pamplona. Sandy ordered 15 Cokes in France because she needed a Super Big Gulp, but they only served up Dixie cups of warm soda. They'd laughed that dogs "Spoke German" in Germany. On a small Greek island, they had dinner plates smashed onto their heads. I later mention this to a Greek friend, and he says, "We only smash plates on the floor... they're making fun of the tourists on that island!" They'd played a card game on the train with people from all different countries, who spoke no

English, but who all grew up with the same games. It sure makes the world a smaller place.

They even toured Dachau, the concentration camp from World War II, for a more serious, historical day. The Jewish prisoners had planted trees and now, over 50 years later, the trees are hauntingly tall, beautiful reminders of the damage man can do to man. They were feeling the intensity of the sorrowful tragedy when they hear "Hey, Mark!" Yep, a friend of Mark's, *at Dachau.*

We laugh the afternoon away. We walk around the cobblestone mall so she can take in the beauty.

That evening we're invited to dinner at Noah and Panda's place, where there are about 10 other friends, including Heater. We play the card game, which Sandy ends up losing, so she does their dishes. She's a real sport and does them... but dries them and secretly puts them back in odd places. Seriously, plates in the shower... they'll be finding forks for weeks.

"So, have you ever done a Ouija board before?" Noah asks as we're all settling in for a few beers.

"No, but I've always been curious," I say, while Sandy shakes her head no.

"Let's try it!" Panda shouts as he grabs it from the game closet.

We gather around as he unfolds the board, filled with numbers and letters, and sets it on the table. There's a little wooden slider with a hole in the top to see which letter it lands on. Six of us touch our fingers to the slider, gently, so no one's pushing it. I can tell Sandy's a little nervous, so she just decides to take notes as we all do it. Noel starts asking questions.

"Is it winter?" The slider moves slowly at first, landing on the word YES and stops. "Are you friendly?" It slides around until YES again.

"Some have been known to lie, so we always take it with a grain of salt at first..." Noah guides us along.

"Who are you visiting?" PAM. *(!)*

"What is your name?" BOB. *(Bob Midnight?)*

The slider keeps moving gently, it's as though each of us is touching it so slightly, like with a feather, and yet, it slides across the board consistently, smoothly, and with confidence. It lets me know Bob is my guardian angel and he's been with me my entire life. He forfeited being born into another reincarnated life to watch over me. I feel a warmth of maybe love, maybe embarrassment, at something so personal being shared in a room full of party pals.

Noel can tell I'm skeptical, so he asks things no one would know. Did I sell my car to go on this trip? YES. What kind of car? TOYOTA CRX. What color? BLUE. What year? 1986.

Okay, in any casual conversation, I might've mentioned selling the car, maybe the color, but I'm sure I haven't mentioned the make or model. I absolutely guarantee that I never, *ever* mentioned the year. It even took me a moment to even remember the year. And Sandy tells me later that she wouldn't have known the right answer even if she *had* been touching it. So, I'm becoming convinced. I like the thought of Bob, and a picture floats into my head of a big man with warm brown eyes, a full salt and pepper beard.

My skin tingles at this visual.

Noah explains that the Other Side likes when we speak to them, and the more we do, the more often they'll choose to speak with us. If I put myself in situations where the opportunity arises, they'll reach out. He asks me if I have a question for the board.

"Is my Dad there?" I ask, my throat tight. There's no response.

I look to Noah for an answer. He gently explains it's probably too soon to speak with someone after their unexpected death. Dad's still figuring things out over there. I stay silent, just nodding. Sandy's been scribbling away on her notepad. We move through the crowd with their questions and while entertaining, I have to say I got the best response by far. We end on a silly note as we ask if they enjoy when we snowboard. The wooden slider hits YES the fastest yet.

As I go to bed, I say my first thank-you prayer to Bob... just in case.

*****

It's the Fourth of July today. Sandy and I put on our snow gear, build a mini three-inch-high snowman, and pose for pictures. We even find a sign at the local pub that's offering a free beer to Americans today. Yippie.

Over the next week, Sandy and I are invited to dinner with friends all the time. Missing-tooth Emmit and his friends, Denny and his roommate, Heater and her friends, and even a couple of parties at Gretzky's, Big Jude's and Skanker's. She's enjoying my vampire lifestyle, up all night, sleep all day.

We end up getting a few new employees at Chico's... one named Dave, a fellow American from Seattle, and his blond and blue girlfriend, Tess, tagged along as a visitor, she's an instant pal for Sandy while I work.

One night, I get invited out by a friend, Rowan, who's a native Māori dancer in a resort show. There's a huge buffet, and Sandy and Tess can come along, all for free. Now, we sisters know there'll definitely be a second plate for us, probably a third, *maybe* even a fourth. We just don't know Tess that well, and to be honest, she's a cute, thin, blonde Brooke Shields. She looks to be an *I'll-just-have-a-salad-thanks-no-dessert-for-me* kind of girl.

The show begins with the men in grass skirts, bare chested, faces fully painted like my tattoo-faced prison pals. The drumming music beats louder and louder as the men perform the Haka, a traditional war dance to intimidate the enemy before a fight. They even stick out their tongues to scare away the enemy.

By our third plate, Tess is putting away the delectable food like a garbage disposal. I'm afraid to get my fingers close to her mouth. We're proud of her as we go plate for plate.

We slurp down mutton, ribs, I think maybe even some rabbit? Then the tribal ladies come on stage to do a stick dance, their chins painted in their ancestry tattoos. Even Sandy's called up on stage to participate. We all go for one last plate of chicken and veggies before dessert. Okay, okay, we then grab two full dessert plates each, end up unbuckling our pants at the table and sneaking a final, and yes *seventh* plate full of nourishment. Most people go back for *seconds*. We classy ladies go back for *sevenths*. *We're in seventh Heaven, ba-dum-dum-tsh!*

I'll bet the free buffet is ended forevermore because of us.

The next few days I click back into my blue and white snowboard that looks like a jack-of-spades playing card. God, it feels great to be snowboarding in New Zealand. The mountains are stunning, the weather's perfect, and our friends are with us. As we meet up with Noah, Panda, and Rick, I hit a steep patch of black ice, stutter to a halt and slip onto my ass and hands, sliding right up to my pals.

I hear Noah tell Sandy, "Man, she sucks!" His voice is filled with disappointment that I'm going to slow them down for the day. They didn't see the patch of ice.

"Great," I whisper to Sandy, "Now I have to prove myself, ugh!"

She just smiles... she knows what I'm capable of. The rest of the day I earn a few pats on the back as I head to black diamonds, jump a few small cliffs, speed through the trees, and give it my all on the half pipe. I think I earn their respect, especially after the boys eat some powder as well.

After just about three weeks, Sandy and I are having to say our goodbyes. Our roommate Chad has laughed at us chopping wood, hiking, making Spam sandwiches, and lighting my steel-toed boots on fire. Hey, walking to work through the snow, the steel toes quickly turn a stinging, painful frozen. I'd quickly wised up, leaning the boots by the fire before I leave home and again, at the restaurant once I arrive at Chico's. One night, the sole of my boot twisted and melted. Too close... now I walk with a slight limp.

I will truly miss this town. My second-to-last night at work, I learn the crew will throw me into the river after my final shift. They do it to everyone on their last night. Yes, that river, that freezing river. The one with the Māori legend that says a Giant was resting on his side while the ground had sunk under his weight to form the meandering river.

So, lo and behold, Chad has a *wet suit* I can borrow. Sandy and I laugh our asses off as I squeeze my big butt into the suit, my Chico's uniform over the top. I feel like a robot, I can barely move my body, it's so thick. We walk the 12 minutes through the snow, and I'm sweating bullets as we climb the stairs to work. Both fireplaces are roaring as we sit down to more meat for breakfast.

"What the hell are you wearing Pammy?" Big Ted's the first to notice.

"Is that a *wetsuit*?" Razza laughs, spitting his soda back out.

Everyone is in hysterics now.

"I heard a rumor about getting thrown into the fjord... I just wanted to be prepared!" I explain.

Spud can't hold back his laughter. "First a Swannie, now a wetsuit! Another first in the history of Queenstown, Pammy!"

I laugh as I slip into the ladies' room and change out of it. There's no way I can wait tables in that suit of armor. When Sawyer Dog sees me, he says it's for the best because they were already cooking up a scheme to strip me down naked and throw me in.

"Red wine," an older gentleman asks as I bartend. "Just put it on my tab." He takes the glass and starts to walk away.

"Sir! Sir! What's your name please?" I yell over the music.

He gives me a stunned eye roll as tells me, shaking his head, continuing to walk into the parting crowd.

"Pam!" Big Ted and Seth pull me aside. "Did you just ask him his name? He's one of our most famous Kiwi's! He's in the news every single night!" *Ooooops.*

Everyone is buying me drinks on my last night. I'm more than buzzed, I'm getting drunk. Someone buys a round of tequila and insists one is for me to "Say Goodbye." I *cannot* stand the smell of tequila. Let's say too many fond memories in an old friend's garage hangout. Haven't had one drop since I was 18. We hold up our glasses and clink. I pretend to drink it, but really throw it past my shoulder onto the floor behind me. Unfortunately for me, it just dribbles down my back and I reek of tequila. The smell makes me want to hurl the rest of the evening.

The crowd dies down, we begin our clean-up process.

I'm grabbed around the waist from behind. *It's time.*

Instantaneously, someone's at each foot, each arm. I'm hoisted up and carried down the cobblestone streets in a big hurrah. Crowds are gathering to see what's going on. One man, whom I don't know, is trying to slip my arm out of my shirt, and even my bra. I'm kicking and screaming. It doesn't feel 'all in good fun' because of this drunk jackass that's trying to get me naked, but everyone else just thinks I'm fighting getting dragged down the pier. He even gropes my breast as he realizes I'm pinned. It's still dark out and my friends swing me out into the great black frigid winter river.

I'm soaring through the air, one arm out, one flopping DD boob out. I splash land, the icy water stabbing at my skin. I get an instant headache against the bitter cold. I kick hard and realize someone must've removed my boots at the last minute. Hands reach down to haul me out. Everyone's laughing. I'm trying to find the groper so I can slap his face. I slap a few friends when I can't find him. It turns out no one knew who he was. What an asshole.

I'm back at the bar, towel drying my hair and reminiscing with everyone. Big hugs from all, they say I won't be forgotten. It's always bittersweet to move on from a town where you truly fit in. Cue the '*Cheers*' theme song.

My last day in Q'town, we meet up with Heater and Skyler for our last cocoa at The Moa. Both girls are heading to Sydney soon, just like us. We can all meet up there in a few weeks.

A new girl was hired at Chico's in my place, and she needed a room—mine. She's just stepping into my exact life. Funny how things work out. I wonder how many times I've stepped into a girl's life as she's moved on, someone I would have the world in common with, yet we're forever separated by just moments in time...

North Island ...rua

Taupo ...rua

- So What'cha Want -
Beastie Boys

Sandy and I take a bus to Christchurch, exploring the town. We head back over the Cook Strait on the ferry. My old buddies Simon and long-red-curly-hair-Jasper drive six hours round trip for pick up in the middle of the night, for just some McDonald's as payment. New Zealand's "Everything" Big Mac includes a beetroot and an over-medium egg... oddly delicious. I think we'll all have in-depth conversations during our trip, but she and I both fall asleep like kindergarteners after a Disneyland day.

We stay with my old roommates Clinton, Nigel, and Andrew of course. We tour the FISH radio station. The Cow is still fun, everyone welcomes us both. We'd like to do more of the tourist thing, but it just won't stop raining. The roomies and I show her the Huka falls and bubbling mud in the rain. I'd hoped for so much more.

We run into Tad and he tells us Sebastian from England and Logi from Iceland had borrowed his car to drive to the South Island. The car broke down and unbelievably, they'd just left the car there. Sebastian told Tad, "It's just travelin,' you know?" This was after Tad had accidentally broken Sebastian's glasses and immediately paid to replace them. I can't even fathom being so selfish and leaving bad memories of myself around the world. What a true shame.

We buy Tad a beer... it's the least we can do.

Jimmy from the FISH radio drives us north. Emerald green hills, volcano and river views help us fall in love with Aotearoa, land of the long white cloud.

Auckland ...*rua*

- How Bizarre -
OMC

With big hugs, he drops us off to meet up with Byron, my fully bearded, shaggy haired, hippy surf buddy from Jeffrey's Bay, South Africa. He now has short brown hair, no beard. He looks like a teenager.

We all hang out for the day around the city, grabbing lunch, coffee, hitting the bookstores. He mentions he's a bible-thumping virgin. *What? This is unexpected.* I guess you never really know someone until you see them in their own habitat. He shows us his photo albums of Africa and he has the same pictures from *The Gunston 500* in Durban, from nearly the same angles as mine. We should've crossed paths but didn't see each other in the city. I truly thank him for introducing me to his friends Brock and Tanner in Cairns and tell him how well they looked after me.

It just won't stop pouring. I'm afraid Sandy's not going to love it here as much as I do. We go to the Kelly Tarlton Sea Life Aquarium and have a few giggles. We meet up with Skyler, who is passing through Auckland too, for a gambling night at the same Sky City Casino that my old Taupo roomies took me to. I see a turtle statue and have to buy the trinket. The New Zealand vibe is winding down, we're getting excited to fly to Sydney in a few days.

At the airport, I have mixed emotions. I love everything about it here... *and yet...* it's got a small-town vibe. Even Auckland, a city of one million. I absolutely feel sad to leave my home away from home, New Zealand.

Kiwis have introduced me to a more profound me. I've been welcomed so warmly, pushed farther into extreme sports (literally, thanks Toad!), and lost myself in such breathtaking natural beauty, usually found only in dreams...

Australia ...bularu

Sydney

- Yolunga -
Dead Can Dance

Sydney. Sandy. Sandy. Sydney.

As soon as we land, we feel the heat. It's still winter in Australia, but after Queenstown, this is like a Hawaiian evening in the fall. Heater flew direct from Q'town, so she and Skylar are already at the airport to pick us up. It feels great for us to see familiar faces even from just a few weeks and days ago. Tonight, there's electricity in the air and we decide to get crazy.

After a street hot dog, and many, many beers, we cross over to the Sydney Opera House. Tonight, it's lit up for a proper Opera Gala. Everyone's dressed to the nines. Long, classy gowns for the ladies, suits or tuxes for the men. The salty sea air adds a touch of the Hamptons to our evening. All of us quickly realize we're in jeans and cute shirts, and, yep, we stand out, and not in a good way. We're much louder than we think, I'll bet. People are staring at us. We cruise in a side door to avoid the extensive line, and lo and behold, we're inside. No tickets needed.

I swear we didn't mean to sneak in.

"I think we accidentally snuck into the Opera House!" I whisper drunkenly.

"Yes, indeed, we bloody did!" Skyler laughs.

"Should we go back out and pay?" Sandy worries aloud.

No way, I think.

"No way!" says Heater. She points to an open door just down a short hallway. We run toward it and barrel through as daintily as we can muster.

The room's filled with amphitheater seating, and we tiptoe *(stomp loudly)* up the steps quickly in case a ticket guy sees us. As if we blend in with this classy crowd. A small orchestra comes onto the stage, just a handful of people, all dressed in black. The main woman is in a velvet floor length, long sleeved, turtleneck dress, with her hair pulled severely back into a tight bun. I squirm in my seat, I'm so hot, maybe from running up the steps, possibly from the change in climate, but her turtleneck just looks so *suffocating*.

She opens her mouth. The audience takes in a collective breath and you can hear a pin drop. Her voice is spectacular a Capella, it melts into a melody on par with golden clouds separating under the sun beams over a tiny village in the Swiss Alps. At an orphanage. On Christmas Eve. The acoustics in the Opera House are simply the best in the world. My eyes fill with tears at the emotional raw beauty of it all.

I turn to Sandy. She's beet red, her hand over her mouth, stifling a guffaw. The kind of laughter where her body's trembling and tears are streaming her cheeks. I can't look away. Now I'm howling. Skyler and Heater take one look at the sister debacle and they crack up.

The audience turns in their seats to dissect us. The violins, violas, and cellos chime in. The entire room fills up like an echo chamber, hauntingly beautiful harmonies, obnoxious Americans and tawdry British hysterics. We should appreciate that we're hearing such exquisiteness. For free.

Cannot. Stop. Giggling.

There's a tiny break in the song. It's clearly not over, but we need to escape. I start clapping. My girls start clapping. The audience has no choice but to clap as well. As we clap, I head for the door, but my steel-toed men's work boots make a heavy clomp on the wooden stairs in ricocheting echoes. The girls follow me out the door. This refined crowd's glad to see us go. I'm mortified that this might've been the singer's debut and she will forever remember our rudeness.

Heading down the hallway, we hear a man's voice, "You there, young ladies! Stop right there, I need to see your tickets!" I glance over my shoulder, still giggling, and see one of the men from the audience who had been giving us the stink eye, now pointing us out to the security guard. *Oh no, is this a real crime?* We did sneak in, but they surely won't believe it was by accident.

"Come on!" Heater yells, and she runs straight out the side door, back past the lingering line of people waiting to show their tickets. We push through the people just fast enough to not get questioned by the guard. We keep running longer than necessary. Slowing down, catching our breath, we end up walking all the way to the edge of Darling Harbor, still laughing our drunken heads off.

"Did we just kick ourselves out of the Opera House?" I ask.

"Yes, but only about 30 seconds before we would've actually been kicked out!" Heater laughs. She's been here for a couple of weeks now. She knows her way around and wants to show us something cool. "Follow me—you're going to love this!"

First, we each grab an ice cream cone and walk around the edge of the harbor with twinkling fairy lights everywhere, wrapped around trees, woven into fences. It really lights up the night as boats are sailing by, the mild ocean breeze in our faces. We follow the crowd and sit along the fountain area.

As soon as it's dark enough, there on the mist of the water sprinklers themselves, they project a movie about the history of the Aborigine tribes. Hundreds of people watch this movie playing on the water droplets, with native didgeridoo music pouring out of overhead speakers. It feels so natural to be sitting here eating ice cream with my sister, ruining the Opera House... yep, our first night together in Australia.

The next few days, Heater shows us around Bondi Beach. It's a cute little happening beach cove, with lots of crowds.

We hang out at a backpacker's place a few days and decide to head north to meet up with a friend of Sandy's she'd met at her ski resort back home. He's a man named Nick who said if she's ever in the Gold Coast to look him up.

Heater's found a job and is flirting with a new bloke, so she's staying in Sydney and will visit us soon.

Sky's up for an adventure. She's here for three months, then back to Queenstown to attend a wedding with Joff.

Byron Bay

- Sly -
Massive Attack

13 hours and we're in Byron Bay near the Gold Coast. What a cute flippin' town. A little forest right on the beach with some open-air cafés, surf boards, and hippies everywhere. We grab a $10 bed at the local hostel and head to the beach for a tan.

"Topless?" I ask. It's the perfect time, my skin's all pasty white after forever in winter, plus if we all do it, I think we'll be braver.

"Ummm..." Sky looks at Sandy and shrugs.

"Ummm..." she looks at me. My top's already off, arms stretched out over head. "Okay then!"

The sun warms our buns like a Cinnabon bakery. Yes, I'm a summer girl. We settle into the warm sand. Unfortunately, we've all gained about 15 pounds from all those kidney and liver pies. Our bellies look Cinnabon-y as well. For shame, but why be fat *and* pasty white? Oh, well, easier to eat more salads in the heat. With a San Tropez tan.

"So, I haven't told you, but Mom gave us each $300 to do something fun!" Sandy says, with a mischievous grin. "What should we do?"

"Why not learn to scuba dive?" Sky points to a sign at a scuba shop right on the water.

"I wanted to learn so badly in South Africa but didn't want to go alone. Yes! Let's do it!" I'm thrilled.

Mom would be proud. When I was three years old, she'd been the only woman in her scuba class of 30 men. It had just opened up to tourism, from exclusively being a military Navy Seal training camp in the early 70's. On one of the dives, the air in her tank had suddenly all leaked out. It just stopped.

Her dive buddy had his own complicated regulator. She signaled to try to share his equipment but couldn't figure it out in time. She had no choice. She swam for air at the surface but couldn't make it in time. Instead, she took in three large lung-fulls of sea water. *She was drowning.* Luckily her teacher yanked her into the boat, saving her with CPR. She had fractured ribs from being thrown onto the boat, but she was lucky to be alive.

"Okay, we'll go ask about it... hey, didn't Mom almost drown?" Sandy asks.

"Yeah, but we'll be fine," I remind her. I'm so excited.

Hours later, we learn it's exactly $300 each to learn to dive. Skyler's going to be our beach babe cheerleader and bow out. So just Sandy, me, and our teacher, Sid, an old surfer with rusty hair, freckles, and an easy smile. It'll be two full class days, a practice day in the pool plus our final test, and the last two days we'll do our first ocean dives.

The first thing Sid tells us in class is that Sandy and I will be diving buddies. Even though he'll be with us, if one of us can't go, the other can't go either. *And blah, blah, blah.* That's all I'm hearing for the moment because I know I'll need to use this information against my own sister like an ace up my sleeve if she chickens out later. *When, not if.*

My sister, she worries a lot. Like A LOT. More than necessary. She's the Good One. I'm a bit more, well, wild and free, to say the least. She loves snowboarding, but only goes on the same run on her favorite mountain. She loves knee boarding behind a boat but won't go outside the wake. While she traveled Europe last year it turned out she called home crying every few days because everything was so different. You see where this diving thing is going.

We learn all about the water pressure, getting the "bends" if you come up too quickly, the equipment, the mathematical measurements of reading the gauges. We have fun in the pool with our practice day. Boom, we pass. I barely do, and Sandy gets a 99%. And next morning comes the first ocean dive, and, yep, that's when the kelp hits the fin.

"Good morning!" Sid waves as we enter the shop. "Ready for your first adventure under the sea?"

Sandy bursts into tears.

Oh, boy, here we go.

"I don't want to go!" she's shaking her head no, chin quivering, tears flowing.

"Oh no, why not? What happened?" Sid asks, walking over for a hug.

"Because we're going to drown!" she sobs. "We're going to DIE down there!"

Sid looks at me confused, like, *what the hell just happened?* I nod, like, *I think I can turn this around...*

"Sandy, you don't have to go," I say gently, hugging her tight. She's clinging to me, blubbering. "This was supposed to be fun. If it's not fun, then we stop, it's okay. Everything is going to be okay..."

She finally spills the beans that she would've felt more comfortable if she'd made 100% on the test. That missing point is life or death. Sid explains a passing grade is a passing grade no matter how many points were down. Hell, mine just squeaked by. I prepare my pretend ace card.

"You don't have to go... I'll just go by myself, right?" I ask, turning to Sid.

"Well, no... just like I told you on the first day," Sid reminds us. "Remember, you're scuba buddies? Can't go without each other..." He lets that last bit just hang out there. I turn and look at Sandy, with a blank look on my face, not saying a word. What she and I both know my look means: *I just spent four days doing math and $300 for NOTHING?*

She stops the tears and musters her courage. *Atta girl.*

"Okay," she says quietly, head hanging down. "Okay, we'll go. But we'd better not die down there or I'm going to be pissed." In Australia *pissed* means *drunk*, so Sid gives us a funny look. But he gets it.

Out on the boat, we pull up next to Julian Rock, a diver's paradise. With our fins on, tanks, weight belts, mask, and wet suit, it all feels pretty awkward at first. I'm told to go first, then Sid, and Sandy last. I sit with my back to the ocean, hold my mask, and fall backwards into the chilly water. I tentatively take my first slow Darth Vader breath, and *presto!* I'm scuba diving. It's technically my second ocean dive because I'd done that resort dive up in the

Whitsunday Islands with my boat friends, but now I really do know what I'm doing. *Well, I passed the test anyway.*

I grab the chain attached to the anchor and lower myself down. At the halfway down mark, I slice my finger open on a sharp metal part sticking out. *Oh no*, I think, as the blood starts pouring out. I hold my breath, take out my regulator and put my finger in my mouth to suck the blood for a second, just wrapping my leg around the chain to hold steady. I watch as a cloud of my blood floats gently away from me. The rhythm is matching the movement of my body in the ocean. It looks so peaceful.

It's that moment, looking through my blood, I first see the shark. It's about a seven-footer, heading toward me, doing that toothy smile.

WELL, FUCK.

Not exactly about the shark, more if Sandy finds out, she will never scuba dive again for the rest of her life. *Okay, maybe a little about the shark.* I just spent $300, four days in class, and fought my way through the obstacle course of Sandy's petrified tears. *I'm not getting out, DAMMIT.*

The shark's getting closer. I put my regulator back in, slowly breathe again, hold a finger over the cut, and climb back up to Sid, who's just below Sandy on the chain. I hold my hand up like a shark fin to my forehead, and Sid gives me the A-OK signal as a reminder that we'd talked about this being a very lively area for all sea creatures. He'd said in class that they won't bother us if we act normal.

Okey dokey. Act normal. Around a shark. Whatever that means.

Sandy's going to be furious if she sees it. I do a full 360 and can't see him anymore. I picture the original '*Jaws*' movie poster. Sid guides us to swim together and follow him. I hang back by Sandy's legs and swim a few paces behind so I can keep looking around for the damn thing. *Is he hungry? Has he eaten this month?*

I glance at Sandy pointing to something—she looks back at me, all excited and happy. She points to a large sea turtle, the size of a car tire, eating an enormous jellyfish, floating right by. With each bite from the turtle's beak, the jellyfish flinches a little bit. Huh, who knew? I keep glancing back over my shoulder.

This continues the entire 45 minutes of our dive... me panicking and looking around constantly for that shark. *Did he get his friends? Do they travel in packs? Was he just a baby and he went to get his—gulp—parents?*

Meanwhile, Sandy's joyfully, blissfully enjoying the dive I forced her to go on. She keeps pointing at giant bright clams, an octopus, an eel, I don't see any of it. She was right. We're going to DIE down here. If anyone gets eaten today, I'd better make sure it's me. I couldn't live with the guilt if I'd forced her to come along and that was it for her.

Once we're on board, I show Sid my bloody finger and ask for a Band-Aid.

"When did you cut it?" he asks, a bit overly concerned if you ask me.

"On the way down the chain," I'm hoping he doesn't add my missing information to the equation.

"Before or after you told me about the shark?" he asks.

GREAT. "Right before," I whisper.

"Shark? You saw a SH-SHARK?" Sandy freezes, mid towel dry.

"Uh... ummm..." time to come clean. "Yes."

"Oh my God! Pam!" Sid shouts, genuinely frantic. "You should've told me you were bleeding! Sharks can smell blood for five miles! They come running like we rang the dinner bell! You were just bleeding into the water for the last hour?"

"Oh, I'm sorry. I did not know that." *Why don't you teach that in class? That would have been considered Primo Information in my book.*

Sandy's sitting down, pale faced, "So, we just went diving... with *a shark*... on our first dive... ever?"

"*...yes...*" I squeak.

She smiles. "Cool," she loves it. She'd had fun while I was the panicked mess.

The next morning, as we head onto our boat, another captain tells us there are whales migrating just a bit past Julian Rock. We agree to take a look before we dive.

We easily spot the pod of gentle ocean giants from their blowholes popping in the distance. We watch in pure happiness as we see one getting closer to our boat.

It seems to be heading straight for us. *Ummmmm, it seems about to ram into us.* Surprisingly, the great beauty lets out a geyser of mist from its blowhole right in front of us. It's moving fast as we hold a collective breath. Nothing we can do now anyway. The whale does a duck dive as it reaches the side of the boat. I could've reached down over the starboard side to stroke her vertebrae as she rounds her back to barely miss us.

"Is she going to tip us over?" Sandy worries to Sid.

"Uh, I don't know, this has never happened to me," he replies calmly, his hand goes subconsciously back to the steering wheel to hold it steady.

We watch her giant spine disappear under the surface. We begin to move to the other side of the deck to see if we can still see her face, but Sandy grabs my arm. We freeze mid-deck and don't even breathe as her tail begins to come out of the water.

Her whale tail is nearly the length of the entire vessel.

"Should I get my camera?" Sandy whispers frantically into my ear.

"Absolutely NOT. This is a once in a lifetime moment. Just breathe it in!" I whisper back.

Time slows down. The wind hushes for a second. The glorious charcoal whale tail comes up over our heads, like a slow-motion wave. The sun holds in between the clouds for a salty summer feel on my skin. A seagull squawks in the distance. The waves crash gently against the side of the cruiser. Still locking arms, Sandy and I turn our faces up to see the full whale tail. Drips of water cascade down onto our faces. I notice a couple of long scratches on the left side of the tip of her tail. If it gets any lower, I could jump up and touch it. A whale tail high-five. With that, in one smooth movement, she uncurls and sinks into the ocean without so much as an extra ripple.

The majestic moment has passed, yet forever etched peacefully into my memory. I'm so wide-eyed, one hundred percent in the moment. We turn to each other in awe, Sid's all smiles, just like us.

"Okay, we're ready to get diving! Let's gear up!" he says.

"RIGHT NOW?" Sandy cries out, "Get in the water with that thing? We're all going to DIE!"

Well, maybe it wasn't such a peaceful, mother nature moment for *all* of us.

After the dive, Sid invites us out for a drink around town. We shower up, invite Skyler, and meet him at the local pub. It's an outdoor café that's owned by none other than the *'Crocodile Dundee'* himself.

Walking up the block, we hear rock music playing live. Once we grab a table, we see the band is made up of all children. They're all about 10 years old, and they're blasting out a kick-ass cover of Ozzy Osbourne's "*Bark at the Moon.*"

I *love* it here.

The sun's setting behind us, not over the water like I'm used to in California. We're on the other side of the world and I just learned to dive and tomorrow's Skyler's birthday. I can't think of a more joyous feeling than great friends, great adventures, and trusting in myself.

I can't sleep all night just thinking about our last few days. It's hard to choose self-care over adventures, but this time, I'm going to sleep in while Sandy takes Skylar up to see the sunrise over the ocean at the light house.

*****

Surfer's Paradise

- Three -
Massive Attack

We find ourselves on the boardwalk in Surfer's Paradise, a few hours north. It's just like Santa Monica's Promenade, or Daytona Beach boardwalk. The city itself is wall-to-wall skyscrapers on the beach, like Miami.

Walking through the restaurants, we settle into Charlie's Café patio area with Sandy's friend Nick. He's a gorgeous blond surfer with a heart of gold. He's a youth pastor, plays in a band, and looks like he should be in a Beach Boys video. He scoops up Sandy in a hug and spins her around, laughing. He's warm to us immediately, like an old pal. His friends, Ken and Barbie, oops, I mean Pete and Michelle, are a couple with two gorgeous butterscotch blondie kids, three-year-old Jade, and nine-month-old Liam. The blond and blue family... who somehow ask us three girls to come and stay at their awesome home as long as we like? Yes, please!

After arriving at their modern, fancy place with a pool overlooking the skyline of Surfer's, Michelle tells us we can help her be a nanny for the kids. No problem. She's just 21 years old and looks like a young Kim Basinger. She's very funny and wants to know all about our travels so far. She seems a little envious, because she met Pete at just 14 and has lived the life of Suzie Housewife thus far. She probably doesn't realize we're all searching for the Pete in our lives, the handsome hubby, good provider, fun and sexy, great home, and healthy kids. You could say we're also envious of her life.

We all settle into the home and spend the next few days lying topless by the pool, eating homemade pumpkin soup, helping change diapers and laughing, laughing, laughing. Michelle's happy to drive us around so we can see the sights. Any time Liam starts to cry, or Jade gets upset, we all start singing, "*Jesus loves me, yes he does...*" We get slower and slower until they both fall asleep. I've got to remember that trick with my future family.

Michelle's sister, Davina, works at the bikini shop on the boardwalk and is a gorgeous, outgoing, blonde as well. We watch Nick's band play and meet his roommate Doug with dark hair and glasses, intellectual-bank-manager-type. His girlfriend Veronica has a brunette bob, who happens to be a COO of a skate company.

Now I feel envious. *These people are my age,* yet they have fully furnished apartments and homes, spouses, children, careers, hobbies. They're grownups. I realize it's all about choices and I'll be getting back to my future soon enough. "This" is a once-in-a-lifetime opportunity that I somehow drew into my life.

Nick has other visitors arriving, too. Leilani's a platinum blond with large fake boobies, with kind of a rough biker side, and her sister, Rachel, a shy, mousy girl. I'm not sure we're each other's first choice, but I'm happy to make new friends.

Sandy has only a few days left before she's home to Los Angeles, and I can tell I'm not even close to heading home yet. She's ready. She misses her own bed, her friends, our family, and even her job. She's ready to be normal again. She wants a Big Gulp and a Del Taco burrito. We spend her last few days really touristying it up. Beach, surfing, pool.

One drizzly day, we all go to Dreamland, the amusement park here on the Gold Coast. Koalas sit slumped over in their trees out in the rain, just keeping their heads down, waiting for it to be over. Luckily, it's a warm tropical rain. Even the kangaroos don't like to be wet. They eat straight from our hands under a tin roof shelter, Jade and Liam laughing as fuzzy mouths tickle their palms. There's a tiger feeding show, including new baby tigers and a majestic white tiger. We go on rides all day and eat junk food. We all bond as we go to a casino to gamble into the night. Perfect day getting to know new pals.

Somehow, Leilani and Rachel get a quick apartment right across the balcony from Nick and Doug. Nick explains to us that Leilani thought she was coming out to fall in love and date him, but he's not really feeling that way about her, so there's a bit of tension now. It's a bit weird that she chose to live so close. And they ask Davina to be a roommate, so she's going to share a bedroom with Rachel. *What could go wrong here?*

*****

It's time to say goodbye to Sandy. We drop her off at the airport in Brisbane for her flight back to Los Angeles. She and I both tear up a little.

"It has literally meant everything to me that you took time out of your life to join me on this adventure," I whisper as I pull her in for a tight hug. "I am truly going to miss you!"

She squeezes me back, saying "This was the most fun I've had, ever!"

When she traveled with Mark, and I with Jay, we could lean on someone else's instincts for a while. I've just had that comfortable break traveling with someone I trust so implicitly. I'll have to be on my best behavior again.

As I watch little Jade waving at Sandy's plane taking off, I really begin to cry now. Sky, Nick, and Michelle all lean in for a group hug. For the first time yet, I feel a pang of homesickness.

Skyler's heading back to New Zealand in a few weeks too. She's been jogging on the beach, and yes, she invites me every morning to join her but I'm a sleeper-inner, not a jogger. She's looking great, helping me to eat healthy. She's truly missing Joff and looking forward to attending that wedding. She's planning her future, leading me to think about mine.

I decide to see about getting a proper job here. I accidentally end up walking into the hottest nightclub in town, Brody's Beachhouse, right on the beach at the boardwalk. As I'm waiting to speak with the manager at the bar, an older, well-dressed, bald man starts flirting with me. I crack flirty jokes right back even though I'm not interested. Turns out it's Brody, the owner, and I get the job. Boom!

I start cocktail waitressing that night, and Michelle says she'll call Nick to tell him I'll spend the night at his place just a couple of blocks away since I'll be getting off at 4:00 a.m.

I get my new uniform shirt and I head out into the crowd and feel the music and just do my thing. It feels like a repeat of the Holy Cow and Chico's. I meet some new co-workers, Kate, with the long sundrenched hair, freckly sun-kissed nose, and enviably fit bod. She's a bartender with an easy going laugh, and she used to live in California. We hit it off. And Mason, the tall, dark haired, shy bouncer from New Zealand. He tells me one of his flatmates is moving out soon, and I can have the room if I need it.

After an exhausting night, I walk the two blocks to Nick's place and ring the intercom.

"Yes?" Nick answers sleepily.

"Hi, it's Pam," I reply.

"Hey, come on up!" he cheerfully buzzes me in.

I walk up the four flights and he's got the door open with a cup of tea started for me.

We chit-chat for a while, sipping tea, and finally he says, "So, what are you doing here at this hour?"

I nearly spit out my tea. "Didn't Michelle call you?"

"No. Well maybe, I was out surfing, grabbed some food and fell asleep..." he tells me.

"Oh my God," I'm mortified to just show up at someone's home so late. "I just got that job at Brody's, and she told me she would call to let you know I'd be sleeping on your couch!"

"Oh, no worries, mate, let me get you some blankets," he says graciously. I realize what a great person he is because he didn't say '*what do you want*' when I rang the buzzer, he just said '*come in.*' Or even at the door, he welcomed me in with tea, no less. True natural warmth and charm. I fall asleep the minute my head hits the pillow, so glad to have new friends I can count on in Surfer's Paradise.

I'll work four nights a week at Brody's, crash at Nick's those days, and spend three at Michelle's on my days off. Perfection.

Kate and I lounge around on top of her high-rise apartment at the rooftop pool, eating cheeses and drinking wines. Her family still lives in Newport Beach, California, and she's been dating a guy who's a pilot and overseas half the time. We agree life's pretty good right now, even if we only have a few bucks in the bank. We go for coffees or sushi rolls until our shifts begin.

One day, I get into a deep chat with Nick's roommate, Doug, about how the world really works. He's a banker and opens my eyes to the fact that *governments don't run the world*. Big central banks do. He says the banks loan money to the corporations, who in turn finance the campaign trails of the political party they want in office. Then, if their candidate wins, those government officials owe huge favors back to their corporate donors, so laws will be swayed to let things slide for said corporation.

"Let's take air pollution," he tells me. "The laws state a company spilling pollution is in trouble, has to pay big fines if they don't clean it up. Let's pretend every company is allowed 100 metric tons. If they're over they get fined, if they're under they get rewarded. Let's say they come in with 10% under, earning their reward. Sounds like the environment's winning here, right?"

He continues as I nod, "But, then lobbyists insist that laws are passed to allow those companies to sell their remaining 10% to the highest bidder. So, the company gets paid twice, once as a reward for the good job, and secondly when they sell their remaining balance to a worse-off company, even a competitor. And of course, there are companies that spew out 150 tons of pollution, but they just buy up everyone else's good ratings, pay the lobbyists to keep this system going, and continue doing whatever they want. The environmental laws are written and passed to sound good to the public, so the public always votes for them, but without understanding the loopholes."

Holy crap, Batman. Wouldn't it be nice if the entire world could continue *breathing air*? Don't these guys have families of their own? Don't they worry about their great-grandkids? It's in ALL industries... food, land, building, infrastructure, medicine, law, whatever, each in its own way.

And who owns the biggest banks? The mysterious Federal Reserve. Right now, in 1996, Bill Gates is listed as the wealthiest man in the world with 18.5 billion dollars. Warren Buffet is second with 15 billion. Yet, what about the names I've heard before? The Rothschilds, the Rockefellers, the Payseurs, the Morgans, the Windsors, etc. You never hear they're secretly worth trillions. *Must be nice!*

What else don't I know? I'm on a mission now, to find out everything I can about EVERYTHING.

*****

To be closer to work, I move in with shy Mason and his roommate Dwayne. I mean Super Dwayne. Super Dwayne is super short, and he's a DJ on a nightly Booze Cruise around the harbor. Each night he dresses up in a Superman outfit *(mini J.R.?)* gets the crowds of bachelorette parties, college grads, frat boys, and party animals going wild with male and female strippers, shots, and club music. He's definitely fun and nice, but also a spaz. We all hit it off.

My new room has just a couple of twin mattresses on the floor, with a dresser in the middle. I pin up some of my sarongs on the wall to make the room my own. It's just a few blocks from the beach and from my job, so it's perfect. Now I can walk home each night after work, plus a block away is a small lake surrounded by tropical frangipani trees... the smell is *heavenly*. I just need one hanger for my dress, and outside in the gutter, I see one hanger lying there. Did I just manifest that? *I just need one Ferrari...*

Sleeping in on my first morning, the birds singing, I hear a knock at the front door. No one's getting it so I get up, hair sideways, no makeup on, just a T-shirt and shorts. I swing open the door and there are six absolutely drop-dead gorgeous body-building movie-star men. *Am I dreaming?* A couple of them have bare chests, their shirts dangling from the waistband of their surf trunks... *Jesus loves me, yes, he does...*

Why, oh why am I not wearing makeup?

"G-day, mate, Dwayne around?" asks the first Greek God.

"Sorry, I just woke up, not sure where he is," I'm not sure where to look.

"Ahhhh, so you're American? We'll just wait here..." he says as they all stroll past me into my new living room. It reminds me of my first day in Taupo at Clinton's house with the cops, the shower, the whole thing. I have a feeling this is going to be another hilarious day in my life.

We all introduce ourselves and they mention they work as strippers on Dwayne's boat each night, and they're coming by to work on their stripping routine with him. Yes, this is going to be hilarious. *And seriously, God, a tiny warning? So I could've brushed my teeth?*

They all take a seat while I start the kettle for some tea and offer lame snacks like pretzels and muffins. They all have not six-packs abs but twelve-packs... one guy even starts to do pushups in the living room. I now, of course, realize they eat kale and sprouts and celery, maybe a fattening cucumber or two. *See, I'm thinking about cucumbers around these nice men minding their own business. Is it hot in here?*

"Nah, thanks, mate," Greek God number five says, turning on the TV.

Just then, Dwayne strolls in with some grocery bags and they all sit around the living room, some on the floor stretching, while they put together tonight's playlist of songs. I sneak into the bathroom, brush my teeth, tuck my hair behind my ear and at least put on some mascara and lip gloss. I decide against using the loo in case while I pee there's a tiny, unpredictable, very rare—*yet very ladylike, I assure you*—toot. These doors are thin and these men are unforgivably hot.

I overhear them discussing 'Should I move my hips like this at that point in the song?' and 'I'll motorboat some Sheila in the middle of this song.' I have a feeling I'm missing quite a show, as I hear them moving around to practice their hip-swivels and pants-tear-off precise moves. I had no idea it was all so technical. One's even discussing his—gulp—*fireman costume.* I throw on my bikini with a tank top and a sarong wrapped low on my hips. I feel much sexier, ready to meet some of these hotties properly. I come out and everyone but Dwayne has already gone.

"Hey roomie!" he shouts in his best American accent.

I smile, "Hi!"

"So, what d'ya think of my male stripper parade?" he grins.

I grin, "They seem really, uh... smart!" I say with a wink.

"Ahhhh, they're all gay Pammy!" he laughs.

Seriously? "Seriously?" I ask.

"They're called the '*Thunder from Down Under*' and they travel around, but lately they've been on my boat. You should come by with some of your friends, take the cruise, what d'ya say, mate?" he asks.

We agree on my next night off, I'll bring Sky and Michelle out for a free cruise. Just then, Mason comes home from his morning run and we have a laugh that I woke up to a house full of strippers. We get to talking and he tells me a bit about his background in New Zealand.

"When my father was a young man, he went out on a small fishing boat with a buddy. Dad reached down into the net to collect a rather large fish, when out of nowhere, came a great white shark and bit off his last three fingers on his right hand. Blood gushed into the water and the shark came back and bumped the little boat and knocked it over. Dad and his friend clambered to climb up onto the flipped boat. His friend did so quickly, but my Dad struggled as he'd just lost his fingers and the blood was making him lose his grip and slip back into the water. The fin started to circle the boat. My Pop tried again. As he was halfway up the side, he kicked his feet and realized the shark had just engulfed his entire lower body. He was inside the shark's mouth, up to his waist."

He continues as I stare, "As he felt his skin pierce, Dad realized the shark was pulling him down. *This was it.* His friend was screaming, waving his arms, trying to punch the shark in the eye. The teeth were searing pain around his torso. Suddenly, the shark loosened his jaws, just for a nanosecond, as if trying to get a better grip to finish him off. Dad kicked his feet against the inside of the shark's mouth, stepped onto the rows of teeth, used whatever fingers he could muster and flung himself out of its mouth and, with his friend's help, he ended up on top of the boat. He'd felt the shark's razor-sharp teeth scrape in slow motion down his entire length of his body with the perfect laceration stripes of blood about an inch apart, from waist to toes. His jeans were shredded. He'd been *inside the shark's mouth.*"

Somehow the Coast Guard found the men on top of the boat just in time, shark still circling, the sea was bright red with blood in the water as a calling card to all nearby carnivores. His father still has the scars-in-stripes decades later.

Oh. My. God.

I haven't even gone pee yet. What a day.

Michelle invites her church friend, Laura, to come on board Super Dwayne's yacht. She's over-the-top shy and blushes as we say hello. What a strange pick, this night will probably be too much for her, debauchery-wise. Now, Skyler and I can handle anything. I've seen her through it all as we cocktailed in Queenstown together, but this chick seems downright timid. She's wearing, like, office work clothes, her blouse has a large bow at the high neck. '*Little House on the Prairie*' all the way.

The yacht is awesome, as expected, and Dwayne welcomes us, dressed as Superman, introducing us to the strippers from the other day at my house. The boat fills up with party people and Dwayne revs up the crowd acting obnoxious but hilarious. House music is blasting and the drinks are flowing. Laura doesn't drink. *Oh, she's fun.* Eye roll. Everyone gets jiggy on the dance floor while others hang back on deck and mingle. Some co-eds are already making out. This is definitely a meat market.

Dave's booming voice through the microphone introduces the strippers one by one. They place a chair in the center of the dance floor while everyone circles around. Greek God number three pulls Laura, of all people, to go sit. Laura! I'm glad it isn't me, but seriously, this is going to be AWKWARD. The stripper, let's call him Triceps, starts slow, waving his long blond hair in her lap, flexing, he slowly rips off his wife-beater tank, exposing his hulking chest. His ripped jeans are popped open and he's out in one second flat, in his G-string, with the elephant trunk in the front. Yep, one of those. As he's bending over giving Laura an eyeful, he's dragging the elephant trunk along her thigh. I look at her expecting to see her squirm, when instead, I see her laughing, throwing her hair around. I poke Michelle.

"Check out Laura!" I yell over the music, laughing to Sky and Michelle.

Our jaws drop as she reaches around his chest for a feel-ski and buries her hand inside his underwear. I can't stop laughing. Laura's now up on her chair, grabbing his head, grinding her hips into his face. She jumps down like a wild animal and pulls his G-string off *with her teeth.* She's on her knees, still dancing, teeth on the skivvies, huge swinging hard dick and balls at eye level and she goes for the full package grab. She lifts her shirt up and squeezes her boobs around him. Even the stripper moves her hands off and tries to dance away from her.

Tears are streaming down my face as I can't stop laughing.

Eeeeeww, dirty stripper dick! Isn't he gay? This is your church friend? What are your regular friends like? Still glad it wasn't me pulled into the chair. Church lady is a wild she-wolf in heat while ***I'm*** *the boring one hanging with the wallflowers? Isn't she sober, for Christ's sake?*

A few more shows throughout the night and Miss Laura doesn't disappoint. At the end of the night, she says it was nice to meet all of us, and can we please keep this quiet as she does have a proper reputation.

No one would believe us anyway.

*****

Skyler and I hang out all day on the beach, we hit Charlie's for lunch, and end up visiting Davina at her swim shop. She tells us every single girl who tries on a suit wears their high heels. So funny. She's getting off work and invites us over to her house she shares with Leilani and Rachel.

As soon as we sit down, Rachel comes in, all wild-eyed, tense and angry, shaking a box of cereal.

"Did you eat my cereal?" she yells at Davina.

"No," Davina looks away, clearly annoyed.

Rachel shakes it at Sky and me. "Did either of you eat my favorite cereal?" she accuses us.

We giggle a bit, "No," I say.

"We've never been here before," Skyler reminds her.

Just then, a tiny red laser dot hits the wall in the kitchen and wiggles a bit. Rachel screams and points to it. "Do you see this? This red dot on the wall?"

"Yes," Sky and I say in unison.

"I have no idea what it is or where it's coming from!" she pulls at her hair. "Do you think it might be a ghost? Or aliens? I see it a hundred times a day! You can see it too? It's driving me crazy!" She bugs out her eyes like she hasn't slept in weeks.

"Are you okay?" I ask. "It just looks like someone's messing with you, like maybe with one of those laser pens or something...?"

"But it happens when no one's home, and all times of the day or night!" She stomps off, clearly a nervous wreck.

Skyler and I snicker, trying to stay quiet. *Damn, we should have stayed at the beach.*

Minutes later, Leilani and Nick arrive from his place. Leilani admits she does the laser pen from Nick's balcony all the time to mess with her own sister. Davina admits to eating all her cereal. Poor girl! She's just trying to have fun in a new country and it's not going very well.

And the next day, Skyler's in the wind. Back to New Zealand, end of an era. Goodbye to the beach days, giggles, babysitting, watching Nick's band play, reminiscing, and sharing adventures as well as our hopes and dreams about the future. We trade some T-shirts and give hugs with promises to cross paths again one day.

It's rare to meet someone halfway around the world that feels like someone you grew up with. She was Gilligan to my Skipper, Jerry to my Tom, Ethel to my Lucy. I'll miss you, my little sidekick...

*****

I'm so sick of men. I've been hit on relentlessly at Brody's by customers, on the beach by a relentless slimy long-haired Brazilian, and then there's my young and handsome, cocky, cold-fish manager at work. Todd. He even sounds like a snot.

"Wipe down table 17, Pam," he barks.

"Okay," I say cheerfully, picking up a towel.

"Pam, wipe down table 17." 60 seconds later. He's clearly a genius, too.

"Oh, I just did when you asked me a moment ago," I smile.

"Do it again," he says, avoiding eye contact.

"Okay," I smile, pick up a rag, and wipe the very same empty table.

I see him coming through the swinging kitchen doors as I happen to be going in, so I hold the door open for him as he passes through, not thanking me or holding it open for me, like a man who was raised right would do. Instead, he yells "Pam, these doors need to remain closed at all times!"

I hold back a chuckle, "I was holding it open for you, Boss." *Or shall I call you Douche?*

Kate notices he hates me from behind the bar.

"What am I doing wrong? I notice he's just a dick to me," I ask.

"Ahhhh, don't worry about him, Pammy. You look just like his ex-wife!" she giggles with me. We'd spent the day together at her pool again.

And then there's a visit from Jimmy from Batam Island. He meets Michelle, Pete, Nick, and all their friends. He and I spend a couple of days up in Noosa Beach where he's visiting his friends. While it's great to reminisce about Asia and have a few beers, a few laughs, and some beach time, the underlying message is: "Why haven't we pashed yet? Are we ever going to pash?" With a "No" and a hug, we part ways as he heads back to Singapore.

I feel defeated. I have strippers in my house each day, but they ignore me. Mason's cool but dorky, Dwayne is always ON, like Jim Carrey. I scrub my face and put on jammies even though it's only 8:00 p.m. *Yessssss, a night at home alone doing nothing. Purrrrfect.*

Through my balcony window, I see the fireworks before I hear them. As I pull the slider open, I smell the smoke. I can hear faint music. I realize I haven't seen fireworks in years abroad. I've missed them on July 4th, on New Year's Eve, even driving by Disneyland in California where they shoot them off each night. Ugh. Stay home? Get all gussied up again to go out?

Just go for a bit... I decide to go out AS IS. I throw a flannel on and some jeans and with my wild curly hair in a frizz mop, not one stitch of makeup on, I'm going to enjoy fireworks and remain invisible.

I walk to the Surfer's Paradise promenade and can't believe it. There's a full carnival in action. There's a makeshift Ferris wheel, other festival rides, a nighttime candle-light flea market, bands playing, food carts, even people walking on stilts and juggling. What is going on here? I grab a cup of tea and sit on the wall at Charlie's to people-watch.

Everyone's drunk as a skunk. All the chicks are smoking hot, parading up and down in the colors of the season, for some reason the vomit-combo of lime green and neon orange. All of them look the same in their tube-tops and mini-skirts and sky-high heels. They seem... *boring.* Parallel-universe-me tonight, my disguise is the face God gave me. The hair God gave me. The freedom in being unseen. *I'm also boring.*

I'm between a nightclub playing techno music, a rock band on the street, and a marching band at the end of the promenade. All the sounds are blurring together when I notice two hippies, drowned out completely as they busk for cash, playing African D'Jembe drums. I watch all the people walk by, not noticing them. One guy is my height with a small goatee, and the other is 6'3" and skinny as a rail, with jet black hair and light eyes, with a tiny soul patch under his chin. We lock eyes for a long moment. They're both dressed like, well, hippies, or like they didn't have enough money to get costumes for the Renaissance Fair.

They decide to pack it in. As they walk by me, the short one yells out to me, "Hey, thanks for listening!"

I smile.

"We're going to get a cuppa, would you like to join us?" he asks.

"Sure," I reply, happy to meet a new group.

We introduce ourselves, the short one is Charlie, and the tall one is Jeremy. Charlie doesn't have any money, so he says goodbye and walks away, leaving Jeremy and me to sit and get to know each other.

"Well, hello there," Jeremy says in a warm, soft-spoken voice. "Sounds like you're a traveler?"

"Yes, but living here now. I work over at Brody's Beachhouse just there..." I point in the direction.

"Are you from America?" he seems genuinely interested. Like I'm not wearing a flannel and have zero eyebrows. Maybe he needs glasses. He blows on his tea.

"Yes, Los Angeles," I tell him, smiling.

"Is that a state?" Jeremy inquires. There's something going on here, I feel *electricity* in the air.

"No, it's a big city in the state of California, on the west coast, like where your Perth would be." I love that he doesn't know. With our movies out around the world, I just assume most people would know all about my country, while it's okay for me not to know about theirs. How humbly naïve of me. "Have you heard of Hollywood? Where they make all the movies? That's a part of LA."

"Ahhhh, yes." His eyes twinkle. "And you also have Yosemite?"

"Yes, in the middle of California." I sip my tea. He asks if I'd been there, so I tell him about one of my trips there, panning for gold, hiking up the falls and feeding deer from my hand with my brothers.

Jeremy tells me he's from a tiny town just west of Byron Bay, and he and Charlie came up for the weekend for the carnival. His sister lives here in Surfer's and they're staying with her.

Just then a friend of his walks up... another American, Sean. He's 37, plays the violin busking for cash, and explains he used to play for the New York Philharmonic when he was just 12 years old. He was a child prodigy. He tells us how it sucks to "peak" at age 12, how his life has gone downhill quickly after that, and now he's homeless in Surfer's, trying to get enough for a ticket back to the USA. He waves to a friend and runs off to join him. My first thought is to "get a new goal." If you've been blessed with such a talent, maybe be a music teacher or something. I meet the most random people.

Jeremy and I finish our tea and decide to walk through the candle-lit flea market. He reaches for my hand. I feel like I'm in a pocket of time. I feel so comfortable with this person I've just met, yet I have butterflies in my stomach. The carnival is going on around us, but it feels like a regular pace and we're in slow motion. The noise, the music, the lights, the stilt walkers are just floating by us. We can hear the ocean waves crashing and smell the salty air. We walk by a table with Native American dream catchers. My skin gives a sudden shiver even though it's hot and humid.

Jeremy takes both of my hands, gently swinging me around to face him. He smiles, looking deep into my eyes, he asks "Do you feel it too?"

"I do," I giggle. I hope he means this electricity between us... not like he'd just walked into a web or something.

He hugs me. "I think this means we should meet for lunch tomorrow, if you're free?" he asks in a gentle Cat Stevens voice.

"I am free..." I smile, a bit coy now.

We agree to meet at Charlie's café, because that's the wall I was sitting on when I first spotted them, and Charlie, his friend, technically introduced us. We hug goodbye and I walk home, realizing I almost stayed home tonight. A whisper told me to '*Just Go*' to the fireworks. Weird. I'm so glad I did.

I sleep so soundly, having been in a trance just walking around with Jeremy. I dream he takes my hand again, and this time we gently fall backwards from the Dream Catcher table and keep falling toward the center of the Earth, toward a gigantic ball of light, a huge ball of pure love. All the while he has his forehead placed gently down on mine and is gazing deeply into my eyes, into my soul. He's smiling. So am I.

I awake to the phone ringing; it drags me from my other-worldly realm but it's okay. It's Heater, from Queenstown. She's been in Sydney this whole time but is breaking up with her boyfriend to travel more, so she's coming my way in the next week or two. First, she invites me to meet her back in Byron Bay where she's meeting another American girl Desiree, from, yep, San Francisco. We can camp on the beach for a few days. Sounds lovely. Second, Mason and Dwayne say of course she can share my room. YAY!

I meet up with Jeremy for lunch at Charlie's. His eyes are an incredible seafoam green in the daylight. They widen at me in soft makeup and my hair twisted to one side. I'm wearing my fancy beige and lace tank top and bright blue sarong over my bikini. He stands to greet me with a hug, ducking away from my floppy straw hat. The day is sunny, warm, and tropical, so we sit outside at the patio along the promenade to people-watch. His drum is on the ground next to his knee. We both order a light salad. Just as smoothly as last night, we get to know each other. Unbelievably, he tells me he dreamt of me, but I feel too shy to tell of my dream. He's in town until tomorrow if I'd like to spend the time with him until he goes home? I absolutely would. I have to work tonight at 8:00, until 5:00 a.m. and I invite him to check out my nightclub.

"That's not my scene at all, but I can walk you home when you're finished if you like?" he tells me.

"At 5:00 a.m.?" I'm shocked.

He smiles. "I'd love to."

We spend the day together on the beach. In the shade of a palm tree, I spread out my sarong while he drums for me. City to the right of me, waves to the left. He tells me he doesn't have a job, he's on the Australian dole for a year while he embraces his painting style. I guess there isn't a stigma with collecting welfare like there is in the USA. He studies Feng Shui and picks up side jobs for people in need.

"What's Feng Shui?" I ask.

"It's an ancient Chinese science about the ebb and flow of energy from all things. How you can help the energy flow smoothly, or how you accidentally block it, with the way you set up your home, decorate, plant your garden, paint color choices... see that building there?" He points to a skyscraper hotel with sharp, triangular balconies at each room. I flip over onto my belly to look. It looks quite modern to me.

"See how the balconies are pointy? They're like knives, jabbing into the hotel next door," he points to the older, run-down, unkempt building next to it. It needs a paint job, a new roof, and even just some landscaping.

Jeremy tells me they can put plants up to block the energy from those balconies, even a couple of small mirrors, forcing the energy back, would be beneficial. He points to a hair salon on the corner, called '*Curl Up and Dye.*' I think it's a catchy name.

"No," Jeremy explains. "Why would you invite death into your business? You'll see, it'll go out of business by next year. No one will understand why." I think back to Asia, how no one pointed with their pointer finger, they opened their palm and showed the way. It was nicer, seemed genuinely warmer.

He shows me another hotel with all glass windows and doors along the front. Very sleek, I think.

"No again," he smiles as he explains. "Look at how that family is trying to find the door, see how the entrance is not clearly marked? How many people just drive by or keep walking by because they can't figure it out? They wanted it to be invisible, well guess what, they *are* invisible but to their customers. But look next door, the hotel with the red awning? Lion statues at each side of the doors—that's a strong entrance. There's even a small fountain with a pond. I'd bet there are koi fish, too. They know what they're doing. A small sight and sound of running water placed just so invites money and prosperity to your door." He smiles. I'm impressed. I'd never heard of it before, but now I can't *not* see it.

After a night of giggling with Kate, and being belittled by my boss Todd, it's 5:00 a.m. and there's Jeremy, leaning by the lamppost, waiting to walk me home, as promised. He casually tucks a frangipani flower behind my ear. How did he know those are my favorite? We hold hands and talk while we walk to my house. He knows I must get some sleep, so he kisses my forehead and invites me to visit his place. I can take the bus and he'll meet me. He's going to hang out with his sister until I wake up, and we can meet for lunch again before his bus home.

I meet him in the afternoon at the coffee place and he tells me about his childhood, asks me about mine. He gives me a gentle kiss on the lips before hopping onto his bus. He's unlike anyone I've ever met. I usually go for the Alpha male, but he's a flower-picking painter in touch with nature. I'm intrigued...

Days later, I arrive into a tiny one-street town, just into the mountains and jungle foresty area to the east of Byron Bay. I've packed so lightly that the bus driver makes a joke about me just heading home from the convenience store. I literally have a plastic grocery bag with a sarong and tank top, a bikini, my toothbrush, and lipstick. I'm going to stay the night and get to know this humble, gentle, spiritual man. He'd already said I can sleep in his room and he can sleep on the couch. Now let's see if we're interested in being more than friends.

He greets me with a huge smile and hug. Jeremy's home is a small bungalow on the edge of a property of a sprawling meadow, next door to a Hare Krishna farm. There's a full-grown male ram in his foresty, orchardy front yard. Yes, a RAM in his front yard. I'm instructed not to look at him because he can get aggressive. *Not weird at all.*

Inside is simple, yet very true-to-being-a-hippie home, with beaded curtains hanging in front of the door, a sarong covering his couch, his drum as a side table, and a partition that separates the bedroom from the living area. He also has a drafting table in the corner with bookshelves behind it. His tiny kitchen has a horizontal mirror across the countertop.

"It's Feng Shui again, Pammy, if you place a mirror where you chop your food, it reflects back that you have double, tricking your mind into being peaceful, that you *always* have enough," he tells me. He points out all the other stuff and how it relates as well. I'm fascinated.

We sneak over to the Krishna farm, along a small lake, so tucked away and private that we decide to skinny dip, even in broad day light. Jeremy's already in the water and turns his back, offering me privacy as I climb in beside him. The crisp, cool water feels delicious against the sweltering heat of Queensland. We laugh, we chat, we swim. As the sun sets behind us, there's a golden glow in the evening air.

A deep voice breaks the silence. A man in an orange robe with a shaved head comes closer and invites us to join him at the Krishna farm. He tells us dinner's being served after prayers in about 15 minutes. We're welcome to come for either, or both, and with a gentle wave and smile, he disappears into the woods. We dress, our skin still soaked, but the heat's drying us rapidly. I'm feeling a bit bashful now that we got caught trespassing, but Jeremy says they're his neighbors, they're cool.

We enter the farm as their prayers are beginning. There must be about 50 people living and working this plantation. They're praying to a small potted tree in the middle of the living room. They chant, sing, and do movements, while Jeremy and I hang back, watching with our heads bowed a bit in respect. Someone hands me a plate piled high with colorful fare. Portions are so generous, like a vegan Claim Jumpers. We chat with the monks while I'm scarfing down their delicious meal, things I've never heard of. Dessert is some sort of ball of dough smothered with rosehip syrup and golden flakes.

With full bellies, we saunter back to Jeremy's home and he drums for a bit at the full moon. Sexy tribal music plays softly while he lights some candles. Sliding next to me on the couch with two glasses of red wine, he doesn't say a word. He places his palm together with mine. There's a palpable energy between our hands. We sit like this, just listening to the music, looking at each other, smiling, sipping our pinot, rubbing our palms. It's *sensual*. I've never had a date like this in my life. Dating in Los Angeles is all about how much money a man spends at dinner, how much cleavage you show. Instantly, the LA dating scene is weird to me and this is how it should be.

Jeremy leans in for a long slow kiss. I get weak in the knees and feel like sitting down. Oh, yeah, I *am* sitting down. He puts his forehead against mine. Kisses my hand. Tucks my hair behind my ear. We kiss for a while. I do the math on how long it's been since I've had sex. *It's not good.* Normally, I'd wait a month of dating someone new, but after realizing this, I go with the flow of the moment. *Can't argue with the numbers, after all.* He pulls me to my feet and we head to his bedroom, section, area, whatever. He slowly undresses me and I him. We explore, make love, and explore again. Candlelight flickers to the beat of the music through the paper-like partition.

"I think I'm falling in love with you, Pammy," he whispers into my ear. We're cuddling, falling in and out of sleep, our breath matching the rhythms of the other. Most LA men would say that to get you *into* the sack, not straight after. He seems deeply genuine to me.

The next morning, he shows me his painting studio behind the house. It's perfect, with bamboo roll down shades everywhere, a comfy forest green couch, another drafting table, a few easels, and stacks of huge paintings leaning against the walls. They're beyond creative, kind of like a weaving '*Jack and the Beanstalk*' vine with thick loops, thin curls, ombre colors... exquisite.

It's a misty morning, so I braid my hair into two rows, like an Indian. Jeremy has no car, no phone, not even a clock. We fill our next couple of days barefooted, eating vegetarian when we're hungry, sleeping when we're tired, making love when the mood strikes, painting, drumming, dancing, swimming, reading to each other, getting to know each other. I feel more like my caveman self. I've never *not* looked at a clock, or run my day by time, or what I *should* be doing. It's lunchtime. Time for lunch. It's time to get up.

Time to get in the shower. Even when I was traveling more laid-back and on my own through Asia and Africa, I still followed the clock.

"So what religion are you?" I ask.

"I believe in Optimism," he tells me, slicing an apple for me.

"What do you mean?" I ask.

"I believe in energy. You can attract good energy with your thoughts or negative energy, depending on how you choose to look at the universe. So, optimism is my only religion. We're all One. I believe in reincarnation like most cultures around the world, I can remember some of my lives. We keep coming back to learn, until we eventually get it right, like Jesus, Mother Theresa, or Gandhi."

I think about Borobudur and the carvings about the nine lives. "So, do you believe in ghosts? Where do you *optimistically* go when you die?"

"I think Heaven is whatever you think it is. That's what it will be for each person. If you're a workaholic at your desk and have a heart attack, don't know you're dead, your Hell would be still working away in your mind's eye. But if you think Heaven is white fluffy clouds and pearly gates, then that's what you get. If you think it's returning to the stars, then that's yours. I truly believe in other dimensions. Do you know how every ghost story in all cultures around the world has the ghost floating about a foot off the ground? Maybe that's just another dimension, one that has their ground one foot above ours."

"Whoa... I love all that stuff. So, what if the stories of Bigfoot, Loch Ness Monster, Mermaids... are all *real*? They just come through what, some portal of time and space? Like the Bermuda Triangle?"

"Well, I think it could be at least a possibility. Maybe they know how to do things we can't do yet. The only thing faster than the speed of light is the speed of *thought*," he sits with me. "You know how people can bend spoons and levitate with just their mind?"

"I guess I've read a little about that," I watch him. *Is he going to bend a friggin' spoon right now?*

"Well, what if ancient Egyptians used the power of mediation along with electromagnetic antigravity sound frequencies to build the pyramids? If the matter in the rocks were manipulated by meditation to just weigh 100 kilos instead of 2000, it makes more sense. It's built with encoded mathematical dimensions of the speed of light hidden in the height and width of the great pyramid. If the entire land on Earth was pushed together, the pyramids are in the center. It's the exact layout needed for some type of a nuclear power plant. It produces energy. It even had mercury pools underneath. All the ancient cultures did this. The 'powers that be' know about this and don't want us to find out we've had the power of free energy all along."

"How do we do it?" I'm floored.

"I can't do it myself yet, but we're all vibrations, it's a matter of matching yours with the universe."

Jeremy goes and searches through his bookshelf, handing me a book called '*The Celestine Prophecy*.' "Have you read this?"

"No, what's it about?" I reply.

"You can take it with you and read it when you can. It's a fun travel adventure that also teaches you about another side of the universe, without all of the man-made religions."

I pack it away with my stuff. I have to leave this afternoon and be back to work at Brody's tonight. Ugh.

"Here's another one, '*Black Elk Speaks*' about America," he hands it over.

We agree that we're truly enjoying each other and that I'll come back for a visit in a few days. It's a $20 bus ticket and a two hour ride each way, so not too bad. I tell him about Heather coming to live with me and he's agreed to come up and meet her, along with my friends and flatmates after that.

The night club life seems strange and shallow after the days I've just had. I just let Todd's arrogance roll off my shoulders. It's still fun, but now it seems surreal instead of Jeremy's place feeling surreal. Kate's happy for me, Michelle as well. My roomies crack man jokes about me getting a boyfriend.

On my next beach day, I start to read *'Black Elk Speaks.'*

I can't put it down, tearing through it in a day. It's about an Indian in the American Dakotas who pretends to be a dummy while government officials want to check into his vision quests and all things Indian. They want to make sure nothing is in place to overthrow the government. He lets them do what they need to do, but laughs at how out of touch with nature, God, even with their own bodies they've become. I'm fascinated by all things native. Growing up in the USA, I never once went to an Indian reservation or a celebration. I'll immediately have to rectify this upon my return, someday.

My first night back, wrapped in Jeremy's arms, I fall asleep to crickets chirping. I have an unbelievably vivid dream of myself as an Indian woman, walking through a marketplace wrapped in a gray and brown feather cape. No one speaks to me. Everyone looks away as I walk by. I can smell the dust of the plains, feel the wind on my face. I remember my childhood as an Indian. It sticks with me as I awaken.

With another drizzly day, I braid my hair to keep the frizz off my face again. I tell Jeremy about my dream.

"That's a past life vision, not a dream, Pam, I'm sure of it," he says gently, yet with conviction.

"Really?" I'm skeptical. "Maybe I'm just dreaming about being an Indian because I'm reading an Indian book, and wearing my hair in braids like an Indian?"

Jeremy smiles and sets a cup of tea down for me. He sits, facing me completely. His aqua green eyes sparkle as if he knows a secret. "Pamela, maybe you're wearing your hair in Indian braids and reading Indian books and dreaming about being an Indian... *because you used to be an Indian.*"

Oh.

"Is that how it works?" I ask, my skin shivering. I rub my goosebumps even though it's warm, humid.

"Yes. And I see you have gooseflesh, that's a sign from your spirit guides. You should pay attention to that whenever it occurs. The other side's trying to tell you something," Jeremy says kindly, sinking into happiness at my learning about his passion. *The night we met my skin tingled in front of the Indian dream catcher table.*

Well, I did say I wanted to learn about everything.

"You know, you and I had a life together," Jeremy hints to me.

"We did? What was it?" I ask.

"I think I'll keep that to myself at the moment," he frowns, lost in thought. "I want you to remember it too. If I just tell you now, I will, at the very least, plant a seed. You'll forever wonder if what you see is truth, or if you just looked in that direction because I told you to. I think the greatest gift I can give you is letting you discover it for yourself."

What a rip-off.

Now it's my turn. "You know, I'm falling in love with you as well, Jeremy."

He smiles. And just like that, I've got a boyfriend again.

We drop our clothes to go for a naked hike to a secret waterfall where we swim, and he tells me this is the purest water I'll probably see in my lifetime. Naked had seemed like a promising idea at the time, until enormous horse flies swarm us with their painful bites. I swat one to make it fly away, but I accidentally hit it *and it still stays on my leg*. I swear he shakes its head at me, as if to say, "*Don't do that again.*" These are Monster Zombie Apocalypse Flies. We rush back to our clothes and keep hiking.

Jeremy teaches me how he sees the universe. He tells me we all have spirit guides that watch over us, some from past lives, some our old ancestors. I tell him about my New Zealand séance and how I met Bob. He loves it. He says I can directly ask them for guidance. I tell him how my instinct yelled to me when my moving truck was stolen, and how, without thinking it over first, I only spoke Spanish in Jeffrey's Bay to those men trying to attack me. He acknowledges those were indeed spirit guide interventions. He tells me there are probably several each day that I'm not aware of, from which food to eat, or which person I'm drawn to. Jeremy reminds me to thank them, and they'll keep visiting.

"Whatever you're thankful for in life, you get more of," he smiles at me.

He tells me to always watch and learn from animals surrounding us. Obviously, creatures hide from earthquakes just before they hit. If I ever see three crow feathers in a short time period, that's a signal from the universe that things are about to change. He reminds me that tribal cultures all over the globe remember how to survive by watching animals, living off the land, and erasing their footprints as they go.

"Overpopulation of this planet is going to become the next biggest problem very rapidly, and it's mostly because of the West," he tells me.

"How so? Most Western families I know have two or three kids, but in the third world, they seem to have seven, eight, nine kids each...?" I remember my conversation with the geologists in Queenstown.

"Yes, but in the truly poorest of those countries, the entire family chops down a few trees to build a small shack to live in. They plant some fruit and vegetable gardens in their front yard, and maybe fish or hunt a few times a week, get fresh water from the stream near their home. In the West, how many people need cars, or a big home, so each get their own room? How many farms are scraping thousands of trees off the planet to make room for their grazing cattle? The chickens, eggs, fruits, veggies are all mass produced, in over-processed soil with zero nutrients. Then half of it gets thrown out once it goes bad. In the West, we water our lawns and fill our swimming pools. We slap on sunscreen and spill oil into the oceans, which in turn ruins the coral, so the fish can't eat and begin to die off. You'll see, it really adds up."

Jeez, good thing we're having fun now.

*****

Over the next couple of weeks, I visit as frequently as I can. Switching between a hippie life with Jeremy, then a few days of beach, Michelle's family, pounding nightclub music, meat markets, strippers, and "normal" city life. It's hilarious, like scrambling to get your six kids off to school, to then rush to peaceful yoga class urgently. When Jeremy comes to visit me, I now see him as out of place in such a busy city. My friends all enjoy him.

Heater/Heather arrives into Byron Bay, so I head down for a girl's vacation, camping on the beach. Jeremy understands, of course. Desiree is a cute pixie of a blond and blue-eyed San Franciscan girl who wears pink romantic lace bodices and flowy flower hippie skirts. She reminds me of Rosanna Arquette. It's great to be back in Byron, where I'd learned to dive just a few months ago. I stop in for a quick hug with Sid at the scuba place. The kid band is still playing their rock music at the local bar. We eat ice cream cones and set up our tents right on the small grassy cliff overlooking the beach. We drink some beers as the sun sets and giggle the night away.

We awaken to the waves crashing, a few surfers getting their wetsuits on and the gorgeous sunrise over the ocean. On the horizon, the sun seems kind of elongated, taking forever, then just pops up, into a perfect circle. After breakfast, Desiree pulls out a baggie of magic mushrooms for a *frying on 'shrooms* kind of day. They taste dry, crunchy, *fungi.*

We stay by our tent and begin doing yoga. Heather whips out her traditional Māori Poi Balls to start practicing like you'd do with nun-chucks. I sip some water and start stretching my legs.

"What do you call your pussy?" Desiree asks as she's doing the tree pose.

"You mean, a name for it, or something?" Heather giggles, giving me a crazy look. "Today I'm calling mine Harry!" I laugh so hard water dribbles out of my mouth. *Hairy.*

"I call it my Fifi..." I say in bursts of laughter.

Desiree changes position to the warrior pose. "Did you know there's an awards ceremony for perfume called the Fifi awards? Hahahaha! I call mine my Girl..."

The crashing waves are starting to look vivid, stretched out, alive. The sand keeps moving in waves as well. It's hitting us. We talk about boys and inevitably, Desiree wants to know how many people we've slept with on our trip. I tell her I've barely gotten a handshake. I like to really get to know someone.

"I'm a slut," she puts up her hand for a high-five. We laugh and high-five, but secretly, I'm hoping we didn't share a water cup last night.

Three ancient, long-haired hippies come over to say hello, beers in hand even though it's about 9:00 a.m. We invite them for a walk on the beach.

"We can't," says Woodstock, who's a skinny, biker type with a Tommy Chong vibe. "We're trippin' on 'shrooms today, mate."

"So are we!" Desiree tells them, giggling. We bond.

They laugh, we laugh, and all begin walking down the beach anyway. Someone hands me 3D mirror-reflecting glasses to wear to enhance my hallucination. I put them on under my sunglasses, and everywhere I look is a trail of rainbows attached to everything.

"We live *inside* a rainbow!" I'm having a profound moment of understanding in the mystical realm.

Everyone looks at me and bursts out laughing. "We sure do, mate!" Woodstock agrees. We each point out seashells, tidepools, crabs, sand, rocks, trees... clouds, the sky, my hair... birds... even an old French fry... it's all fascinating.

They show us around their town, telling us they're homeless and fry on mushrooms every single day. They always scrape up enough money for beers though. We all just chat for the next hour or two, sitting in the palm tree shade, moving through the park, ending up at our tents again. They wish us farewell and leave as we girls put on some South American pan pipe music and just watch the waves until the end of the day.

I head over to Jeremy's. Heather and Desiree will come up to Surfer's to stay with me in a few days.

I tell him about the mushrooms. He's disappointed in me. *Yikes.* I'd just assumed that as a hippie, he'd also be into smoking pot or eating mushrooms now and then.

"Pammy, the purer we can keep our body, it keeps us in tune with the universe. Eating just fruit and veggies, using natural soaps and things, that allows us to be more receptive to the other side and what they're trying to tell us. Nature is perfect. *We* are perfect. If we can keep it that way..."

Oh well. I did have fun, but maybe he's right. I barely drink anymore already, might as well see what I've got to offer myself at my best. What else don't I know?

At breakfast, Jeremy tells me how he believes in aliens. He says he saw a UFO with his friends when he was 15. The group of 10 hiked up to the teen hangout mountain peak, hanging around for a few hours. Out of nowhere, there was a slow moving, stealthy, quiet *UFO*. He says it was as large as a football field with lights all around the edges and hovered above the town for 30 minutes. He and his friends were frozen in awe and no one spoke. As suddenly as it had arrived, it took off in a shot of what looked like stretched lightning and disappeared in a nanosecond. Afterwards, they all ran home and told their families. Some were believed, some were not.

"How come it's hard for some people to acknowledge alien life? There's evidence all over this planet," he wonders aloud. "Look at the carvings from different eras, all over the world, that have some form of a creature wearing what looks to be an astronaut helmet flying through the sky. Ancients bowing down to a flying disk? The one called the Dendera lightbulbs? The guy in the rocket ship, with a breathing tube and his foot on the gas pedal? There's evidence *everywhere.* Ancient Maya, ancient Egypt, the Inca and Aztec, Aborigine art, all over the Middle East, China, Russia, even the USA."

I ask him to tell me more. "*We* are the aliens," he smiles, shrugging, like he knows I'll think he's crazy.

"If we went into the universe for 1000 years looking for life on other planets, our muscles would atrophy from no gravity, our heads would be giant. Our pupils would dilate 100 percent to allow us to see in total darkness, and our eyes would be huge from not ever seeing the sun. Just look at how we have slime in our mouths, in our eyes, like they put into the alien movies. If we cut our arms, the skin naturally tries to adhere back and fix itself. It's crazy, the clues that are right in front of everyone. We all just need to slow down and breathe, meditate to see the truth."

He continues, holding his hand up in front of his face "Look at our fingers and toes, they move independently of one another." He looks like ET, moving his fingers around. *Are you sure you're not on 'shrooms just now, mate?*

"They're coming more often. You'll begin to see more and more movies and television shows about aliens, books, and witnesses before the big reveal. The governments around the world will have no choice but to admit they knew all along and were trying to cover this up."

"What are they doing here? What do you think they want?" I ask.

"Well, I think they want our gold. Not in the way we do, it's not because it's shiny or worth money, it's because it's the best metal for electric conductivity with the highest resistance to corrosion. That means even though the other metals might also be good, they don't last forever. Gold does. Did you know in the 70's NASA sent out a golden record into space with music and pictures on it, all about the Earth?"

"What? No, that's amazing!"

"Yes, they can still communicate with it, the Voyager or something. I think the aliens helped us through that piece of history with the missing link of how we jumped from Neanderthals to the humans we are now. Maybe they designed our DNA. I think they created us as slaves to mine for the gold, then we became advanced enough that they decided to sit back and watch this journey we're on, see if we can push past this lower vibe and get it together. I think they monitor our wars."

I tell him about the secret Area 51 in Nevada that we're told does not exist, but you'll get shot if you get too close or ask too many questions about it. *Yep, nothing to see there.* I tell him about Roswell, New Mexico, and the crash in 1947. Witnesses said the *Unidentified Object* that was *Flying* broke into hundreds of pieces, and each piece was a light metal, like tin-foil, but when you crumpled it up, *it un-crumpled by itself until it was flat again* after a few seconds. The military quickly grabbed up each piece, even illegally searched people's homes to snatch up every trace. They said it was a weather balloon. *Then why search homes illegally?*

Within the following two weeks of the Roswell crash, NASA and the CIA were created. And wouldn't you think the top-level secret government clearance is for nuclear war military secrets? Nope, it's not. The top clearance is for UFOs.

Just a weird coincidence, I'm sure.

*****

I decide I love it here and with my visa about to expire, I spend the day in Brisbane extending it at the Embassy for $200. They stamp my visa, then I go to the travel agency to buy my ticket out, I decide on Bangkok, Thailand. Unfortunately, everything is fully booked until the day *after* my new visa expires. I grab the last ticket left for Asia and only once I'm at the Embassy again I realize it's the end of the working day for everyone else. I head the two and a half hours back to Surfer's Paradise to begin my shift at Brody's. I work all night and am the first back in line at the Embassy in Brissy to see about extending my visa just one more day to accommodate my flight.

The chubby, sweaty clerk behind the counter frowns at me.

"It's going to cost you $200 to extend for the extra day," she says smugly.

"I understand," I say humbly. "Is there *anything* I can do to waive that since I was just here last night? It's just the one day to catch my flight?"

"No. No. No," she says. She grabs my passport to look at my visa again. She's mouthing the words as she reads. Even I know that the different colors mean different things. *She must be brand new here.* "No, there's simply no way around it. There's nothing we can do for you here. That will be $200."

"May I please speak to..." I grin, sheepishly, my voice dripping with honey, "... *ANYONE but you?*" I shrug my shoulders. I'm positive that didn't come out right. She sighs, but to my surprise, she grins back, gets up and brings over the manager.

"Yes, I can waive the fee since you're back straight away this morning!" he tells me. Yippie!

Maybe I shouldn't have worn my work shirt to extend my tourist visa. *Allegedly...*

Heather arrives and gets along beautifully with everyone. My pals all prefer "Heater" again, so we're back to that. Desiree's meeting other friends. We stay up late talking the first night and she points out our cool new address.

"Sunrise Boulevard, Surfer's Paradise, the Gold Coast, Australia, I LOVE it!" She also tells me about her boyfriend in Sydney. He was kind of a tattoo biker type, lots of fun, but she knew she wasn't going to get serious with him.

"Interestingly, his father's in prison the rest of his life. I got to meet him because he gets an ankle bracelet pass for two weekends a year to see his family." She tells me, her curly brown hair up in a bun, unpacking her backpack, making a pile on her bed.

"Why? What'd he do?" I go up on one elbow to face her.

"He worked for the airports as a landing guy, you know the ones that hold the little orange lights and wave a plane into the right spot? He told me that years ago, he was out at a hidden military airport that no one ever acknowledges in the middle of the desert, when he saw a few unmarked private jets land and pull directly into a hangar. About 12 men got out, the most powerful people who run the world, they're called the Illuminati. They run the banks, tell Presidents and Prime Ministers what to do around the world. He told me a bunch of names, but all I remember is Rupert Murdock, George Soros, a Rothschild, and a Rockefeller. He thought one was even from the royal family. He went to report it and they arrested him, telling him he did NOT see any planes land, even though they were still sitting there in the hangar. They had a long meeting and flew off again. He's been in jail ever since, and if he tells them they're right, that he never saw it, they give him visitations with his family."

WHAT? HOLY CRAP!

Next morning, Heater loves the '*Thunder from Down Under*' living room parade, of course. Giggling, we walk down to the beach. We're in the shade as she puts mega sunblock on, long sleeved thin shirt and a big floppy hat so she doesn't get burned. "*Slip! Slop! Slap!*" is the Aussie's logo for sun protection. Slip on a shirt, slop on some sunblock, and slap on a hat. That British skin. Poor thing had to cover the entire camping trip in Byron, too. I take my skin for granted.

I lie on my sarong, cracking open the first page of '*The Celestine Prophecy*.'

"Will you read aloud to me?" Heater asks as she curls up on her beach towel.

The story unfolds and we love it. In the jungles of South America, a restless guy's looking for a mysterious manuscript with the answers to life. As he finds clues and meets the right people to help guide him to forward his journey, he finds the chapters of the book. Looking for coincidences, following your instinct, understanding connections between people, and the knowledge to know more. We take turns reading for ages.

"Hey, it's kind of like us," Heater muses. She grabs the book to read to me for a while.

"How so?" My eyes close while she reads in her sweet, posh English accent.

"Well, we were restless and followed our gut telling us to travel, we met the right people along the way, and now you're my closest friend on the other side of the planet. We get handed a book like this and we're reading to know more and more about everything!" she nails it, perfectly.

"Wow, how funny we were both drawn to the same place as each other in the exact moment in time. Yet we were raised so differently, so far apart. I feel closer to you now than lots of friends back home that only know me from before I left. They wouldn't get half of the things I've seen and how it's all changed me to my core." I feel understood, by Heater, by this author.

She rolls a large joint, lights up and passes it to me. I take a puff and hold it for a moment. I still can't take a big breath or I'll cough for days. I hadn't been high in months, since Taupo, since my skydive. I understand that I was abusing it, letting it take over my life, my goals. But today, I use it to open my mind, so I'm okay with once in a blue moon. I feel balanced.

We walk home and chat with the roomies for a moment before they dash off. Heater puts on some tribal chanting music and we lie on our beds, lost in meditation.

I let my mind wander and I start to think of the people back home. A face flashes in my mind's eye and suddenly my skin tingles. I notice it this time because of what Jeremy had said. I think of more faces, more loved ones. One at a time, and most of the time I get a good vibe. Then one comes along that's taken advantage of my friendship, my finances, my kindness. My skin goes cold and clammy, my stomach in a subtle knot. I reverse and think of someone good and the tingles come right back.

I feel like I'm getting messages from the other side. For the next hour, I let my mind and body tell me about everyone I know, and I get a much clearer understanding of how they fit into my life and I into theirs. My ego isn't involved in choosing if we stay friends or if I feel like forgiving, I just let it all float on, tingles and clammys, telling me what to do. Who is important to me, to my soul...

*****

Today's my 26th birthday! Michelle's planned a party for me at the local pool area at Nick's complex. Jeremy, my roomies Mason and Dwayne, Heater, Desiree, Davina, Michelle, Peter, and the kids, Rachel and Leilani, and Kate from Brody's are all in attendance. Nick and his band are going to play a few songs. Michelle's baked a caterpillar cake complete with rainbow frosting and M&M's all over it, even sparkler candles as antennae. My own Mom couldn't have done a better job.

Everyone brings me a small something, a candy bar, a whale tail necklace, a Calvin Klein T-shirt, some new tea. Everyone except Jeremy. I guess I thought he might pick up that little $10 skirt he'd said was cute at the flea market. Or he could at least paint me a picture. Hell, he's an artist, no?

I look at Jeremy, his seafoam green eyes are shimmering as he gets us both a slice of cake. He smiles.

"What does Calvin Klein mean?" he asks softly, as he hands me a plastic fork on this sunny summer day in November.

Seriously? Again, I'd just assumed that if you weren't living in the Antarctic somewhere, you'd know about a world-wide brand. It doesn't matter, it's shallow. I'm glad he doesn't know.

The very next night I get fired from Brody's Beachhouse by Dickhead Todd. I need the money, but so very glad not to have to bow down to his arrogance anymore. My flatmate Mason's truly sad for me. He doesn't like Todd either. Jeremy had gone home, otherwise we could hang out. Kate's bummed for me. As we lie out by her pool the next afternoon, she asks what happened.

"He just kept at me about wiping down a certain table, and finally I told him I got it, I can handle the job he hired me for." I tell her, popping a grape into my mouth, getting up for a quick dip, already half dry from the heat in the three-second walk back to my lounge chair.

"Oh, no, no, Pammy, here in 'Straya, we do whatever they say immediately. Like if he asks me to get him a glass of water even though I'm serving someone a beer, I'll stop filling the beer midway, put the glass down, and go get his glass. Then he'll feel like the dickhead and say, 'no, no, please finish what you were doing' and get it himself. After a while of doing that, they leave you alone." Ohhhhh, oooops, that makes me the arrogant one, asking to be left alone. I'd always finished what I was doing, *then* did what they asked. *It was just so obvious.* I follow Kate over to Brody's and sit at the bar to chat with my old friends, when Brody himself walks in, sees me sitting there and makes a beeline.

"Hey, Miss Pammy, how ya goin'?" He says with a wink.

"Hey Brody. Just grumbling to friends about missing working here," I tell him.

"You don't work here anymore?" He looks truly concerned. "Why not?"

"Todd and I had a... *disagreement*," I tell him.

"Oh yeah? What about?" he asks.

"He disagreed with me on how big of an asshole he is," I can't help a laugh.

Brody leans in and whispers in my ear, "I've just fired Todd. Start your shift tonight?"

I wrap my arms around him and kiss his bald head. "Yes!"

He loves it and buys my beer and introduces me to the new manager, Elliot.

As soon as my shift begins, in walks Todd to collect his last paycheck. He looks at me, his eyes widen in disbelief. I smile and say hello. His face reddens in embarrassment, he's furious as he looks away, quickening his pace. *That's right, off you go, asshole.*

*****

Back at Jeremy's, he takes me to meet his mum for her birthday. We hitchhike to her lovely ranch on the top of a picturesque hill. Wild peacocks and baby goats roam her home along a wrap-around-the-entire-house porch. She's warm and beautiful and so proud of her son. She seems so... *normal*, not as much a hippie as I'd expected. He shows me around his favorite old spots on the ranch and we go out to dinner. Jeremy's been so poor that this is our official first date it seems. Oh, wait, his mum pays for all of us. We spend the night and head back the next morning, all disheveled.

As we're sticking out thumbs to hitch back, a car full of college frat boys speed past, yelling irately at us, "Ferals!"

"What's 'ferals'?" I look up at Jeremy.

"It means unkempt, like a stray cat, out in the wild..." he says, eying me to see if it hurts my feelings.

"Yes, that sounds about right this morning!" I laugh. I hadn't realized someone could be mad at Jeremy, ever. He's so '*Gandhi*,' calm and peaceful about everything.

Back at his house, we grab some watermelon and head to our swimming spot at the Krishna farm. We laugh as I spit watermelon seeds all over him, grabbing slice after slice, before we dip in the cool water.

It's only once we're back inside his home that I feel the stomach pangs. They're getting sharper. Oh, no. *Too much watermelon.* It's happening, I'm going to have *diarrhea.* On a date. At my new boyfriend's house—that's a 300 square foot bungalow. *Zero buffer zone.*

I excuse myself to the loo. I can't help my body symphony. I make some didgeridoo noises of my own. I add some feral cat in the jungle, the tuba, even a ships horn, followed by a kazoo. I hope Jeremy got the hint and went into the artist studio around the back and played some genuine music. Finally, I bashfully come out.

Jeremy's just sitting on the couch exactly one foot away from the bathroom door staring at the wall. *Nope, couldn't have gone for a walkabout? How about that drumming you always love to do at random times? Not even running water to pretend to be doing the dishes?* It's the moment I comprehend that a TV is a necessity in my life. It's fun running around with someone so in touch with nature, *but c'mon, an old rerun of 'Cheers' would've been great background noise, yes?*

It turns out today's also Thanksgiving. Jeremy asks me all about it and we have a vegan celebration to share together. It's my first bit of nostalgia in ages. Could I really stay in Australia, get serious with Jeremy... *forever?* I know my mom is pulling out all the stops, all the family will be there. And here I am with tummy troubles and, I guess, a pile of zucchini and eggplant, or whatever this is.

*****

Heater and I spend many an afternoon at the beach reading to one another. '*The Celestine Prophecy*' is like our own personal guide. We share our recent dreams, then the chapter of the day is deciphering dreams. Days later, we bicker a moment, the next chapter is about why people argue. Another chapter is about eating as cleanly as possible to be receptive to the signs in your life, just after we'd splurged on a bad pizza and decided not to waste money on stuff that makes us feel so groggy.

Seeing how science tells us most humans use about 10% of their brain power, I'd always believed that Jesus himself had possibly been able to access 100% of his brain, maybe that's why he could walk on water, change water into wine, heal people. The end of this book ties in with Jesus in such a way that I found a love for him even more. He was saying not to worship him, *but God is inside each of us.*

Our thoughts. Our feelings. WE are the Universe. And we're moving it forward.

*****

Finally, Christmas is upon us again. I celebrate at Michelle and Peter's. Their entire family's there, little Jade opening her toys, squealing with joy at each one, Liam laughing and clapping. Davina cracking us all up, more sisters, friends, along with their parents and grandparents. I feel honored to be a part of their intimate family gathering. They even swap a few gifts with me.

I see Jeremy later in the day and I have him unwrap the statue of two people in an embrace from South America. I'd bought it for him at the nighttime flea market. He smiles, tells me how much he loves it and hugs me. He has no gift for me. Again. Couldn't even make up a drum song for me? *Really?*

Everyone has the holiday vibe, vacationing from work. I receive two different surf invitations. One afternoon I surf with Nick on his borrowed long board, as I still remember my training in Africa. I do okay. Next day Super Dwayne and Mason take me out again, we're all on short boards. I paddle out but fall every time. I still love it and feel like a cool surfer chick in Surfer's Paradise.

New Year's Eve I'm back in Byron Bay with Jeremy. We camp on the beach, in the same spot as I did a few months back with Heater. She'd gone back to Sydney to visit her boyfriend. I'm so happy to be here. There's a New Year's festival, everyone in costumes, belly dancing, drumming, painting, palm reading, tarot cards, magic tricks, meditation, massages, reiki booths, jewelry, even pixie dust... *okay, I made up that last one.*

We come across a booth with large, smooth stones about the size of a russet potato, with faint stripes across them. Jeremy tells me they're highly spiritual rocks from the Ganges River in India, and if we hold them for just a moment, we can feel their energy.

I pick one up and it feels unusually heavy, and very cool to the touch. I copy him and hold it steady, focusing on my hand, my arm. My skin starts tingling very softly. I feel the goosebumps begin to form, but not all at once. First the hairs on my wrist stand up, then my forearm, my elbow, the energy feels warm, even though the stone is cool. The blood's flowing through my arm, I faintly imagine the exact second all my capillaries are opening. My heart beats a little faster. As I put the stone down, the feeling stays with me for a few moments, then slowly melts back down into my hand, then gone.

We meet up with Felix, another of Jeremy's hippie friends. He's mid-40's, a rainbow tie-dyed shirt on, with his long gray hair in a ponytail. He's on his 25th day of fasting. I can't understand people like this, I've never missed a meal in my entire life. I wish I had the drive, desire, and stamina, but I just don't want to. I buy a frozen chocolate covered banana while we're walking. *Sorry, mate.*

Jeremy walks ahead, catching up with a couple of friends, so Felix and I talk about politics, and he tells me what he's learned of the Persian Gulf War.

"When a country wants to remove large sums of money from the USA, they need six months advanced written notice. The only three reasons you can't get your money is if there's a government coup, or if you go to war, or if you get invaded. In any case, you then can't ask for your money again for another decade," Felix pauses, sipping from his lemon water.

"So, Kuwait gave written notice to remove billions from your USA banks. Three days before the six months was up, Saddam Hussein *suddenly* invaded Kuwait. Can't get their money. It looks like your US government asked Saddam to go in and start this war, but then once he got into Kuwait, he reneged on the deal and refused to leave. Hey, you guys just handed him a free country! So, your government decided they could look like the hero and kick him back out. Win. Win. You don't have to give up the money, plus everyone will think you helped out the little guy!" he points to his brain, like it's a genius move.

"Is this true?" I ask. I had many friends in the military, some ended up going over to fight. Are we truly just pawns in this grand game of chess? Or make that a grand game of *RISK*.

"I love looking at world politics and finding truth," he says. "You just can't watch the news to learn it, the news is owned by the bad guys too. There are underground channels that you find when you look in the right direction. The greatest scam on this Earth is whenever someone stumbles onto a kernel of truth, 'they' will shout 'Conspiracy Theory!' The entire term 'Conspiracy Theory' was invented by the government. Everyone stops looking after that. They use violent events or big court cases planned by world players. They're called False Flags."

Felix continues, "They did WWI, WWII, obviously JFK's assassination. In the Zapruder film, you can see the few moments after Kennedy was shot the first time in the neck, the driver kept looking back to see if he was dead. When he realizes he's just wounded, *it's his job to stop the car fully* in front of an open gutter. It looks like the police motorcycles close in on the vehicle, when in reality, it's that moment the driver actually brings the car to a full stop for that fatal shot, so that the *coppers rush up on them*. The film's been doctored to bring the car up to normal speed... but there are pictures of a little girl and a cop looking to the gutter at the gunshot, not toward the grassy knoll..."

I'll need to see that film again. 'Back and to the left' becomes crystal clear.

"There were eight gunmen placed in different spots, but the final shot that killed him came from someone hiding in the nearby gutter," he pauses as my jaw drops. "You know the actor Woody Harrelson? His father was thick in the Mafia... he was definitely involved."

"Wait, Woody from '*Cheers*'? You're kidding with me now!" That can't be right. *I'm not sure what's what anymore.*

"Plus, as Lee Harvey Oswald's getting arrested, he's yelling loudly 'I do NOT resist arrest!' over and over. Isn't that a strange thing to be yelling? Unless you know the 'plan' calls for *someone* to be a patsy, and in that moment, you realize it's *YOU*? Strange he was shot before testifying, right?"

"Well, I can't believe there's anyone that still believes the official story, the details have always been... *bizarre*. Sloppy, even," I agree.

He nods. "But there's even more subtle stuff... stock market manipulation, vaccines filled with toxins, deals with big pharmaceutical companies, drug trafficking, even human trafficking. I heard there's *a city* of kidnapped kids under your Getty Museum in Los Angeles."

What?! What are you even talking about?!

"There's even a sex island in your Caribbean where the Elite fly out to get with kids. They drink the blood, it's called adrenochrome. It's like the fountain of youth, I guess. Terrifying. They've been doing mind control for years, it's called '*MK Ultra*,' with LSD and fear techniques since the 50's. If you really look into it, down the rabbit hole you go."

How does one "look into" this stuff? At the Library? Interview a bad guy?

"They have something new called '*Agenda 21*', where they have a plan to abolish private property, make pedophilia normal, create a massive immigration problem around the world, take away guns, all by 2021. If they control all the media, they can tell you whatever they want. Information control IS mind control. The '*Emperor's New Clothes*' for sure, mate, look into it, I tell you."

What does he think I have? Some sort of Pocket Encyclopedia?

"Hollywood is tied into it as well. They'll hire a big director to show pretend bombs dropping to induce wars. Then they finance both sides of the wars. It's disgusting. If we're all fighting each other, we're too distracted to fight *them*. The Illuminati, the Bilderbergers, Freemasons, the Trilateral Commission. Secret Societies go by many names. 'Skull and Bones' at Yale University has launched how many Presidents, Vice Presidents, even CIA and FBI leaders?"

Way to pull back the curtain. Funny, since I am in Oz...

"The Royal Families are the worst. They're heavily tied with the Nazi party. And don't get me started on the church. Did you know there was even a Pope Joan? If you look way back, they've erased her from the history books. But she was there, I assure you!" He waves at another friend and runs off.

WHO WAS THAT GUY? My brain *hurts*... and I need a *shower*. Is the world really so... *ugly? Ignorance has been bliss.*

I walk over for a hug from Jeremy.

I meet one of Jeremy's cousins, Leif, and he invites us to a meditation for the Eckancar religion that I've never heard of. I'm just happy to clear my mind from Felix. We go with him into a small tent and he explains it.

"Have you ever had a small squeak tone for a consistent moment in your ear?" Leif asks me.

"Yeah, sure, once in a great while," I nod.

"Well, Eckancar teaches that might be God trying to speak to you. If only you were on the same frequency, you could hear him and understand what He's teaching you. Here, if we sound out our meditations in a deep breath, and while you let it out, slowly say 'Huuuuuuu...' and keep repeating," he tells us.

We close our eyes, take deep breathes and all synchronize our sounds of "Huuuuuuu..."

I *want* to feel a connection, I really do, but I feel foolish. We do this for 30 MINUTES. Ugh. *Huuuuuugh Hefner. Huuuuuugh Grant.* I find my mind wandering to the noisy party outside. *How long do we have to do this? Maybe God doesn't have much to say right now? Oh, hey, I* ***can*** *hear God. He wants me to go back to the party and have some fun.* This. Is. Not. For. Me. Finally, we say goodbye with a hug, while mentally I can check-an-car Eckancar off the religious To-Do list.

It's still warm out as we approach midnight, and we hear music coming toward us. It's bringing my vibe back to happy. At the center of this party, I spy a very short, very tan, old man with a full head of curly salt-and-pepper hair. He's sporting a wide grin, carrying a scepter, and he's dressed in only a silver crown, tight shiny silver bicycle shorts, and a cape. He seems to be leading a dancing parade of faeries, lizards, drummers, ghosts, belly dancers, animals, hippies, and jesters through the streets. Hundreds are following him. My hips move as I join him and we dance arm and arm leading the entire parade, with Jeremy drumming a few people back.

I come face to face with Woodstock, the hippie from before. I smile and wave hello, but he doesn't remember me already. Someone hands me a tambourine and I play as if I'd ever played before. Everyone suddenly has sparklers, including me. We're marching the streets in costumes, dancing, hooting, and hollering like a gay-parade-conga-line for the last hour.

The parade moves back onto the beach, a circle of tiki torches lights the way, and someone has brought an Asian Gong onto the sand. It's surreal as the gong sounds at midnight, then the waves crash onto the beach, *gong, wave, gong, wave*. Then, on the 12th hit, the sky explodes in cascading fireworks.

This is simply the most fun New Year's Eve party, ever. Maybe because it's warm and tropical and I've left my shoes behind hours ago, dancing through the streets. Maybe because I'm on the beach. Maybe it's really the people. They shared a moment with me, a stranger in a strange land.

Jeremy leans in for a kiss to begin the new year...

*****

Michelle, Peter, the kids, Leilani, Rachel, Nick, and I are spending the day at the beach. Nick's telling us about his trip with the church last week to the Solomon Islands. His group traveled to the small villages and delivered much needed supplies. They got to know the local tribal elders while learning about their culture. He said the native kids were fascinated by the puppet show about Jesus. They'd squealed with delight and awe as Nick pulled out a few little dark-skinned felt puppets.

Nick's teammate, a woman named Julia, had brought shipments of toilet paper as gifts to the town elders. I feel Nick cringe as he tells us how embarrassed he was at her handing out each roll. It sends the message that how they've been living up until they were lucky enough to meet Julia was clearly the "wrong" way, and our way is the correct way.

He even has a mysterious story about how an elder took him hiking way back into the jungle where tourists don't normally go. They arrived at a thatched roof hut filled with human skulls piled high in the middle of the dirt floor. One skull was upside down near the entrance. Nick was told this was a spiritual place where they honor their dead. And that some nights, with enough tribal dancing, prayers and spiritual energy, the skulls come *alive*. The one skull that was upside down near the door had "walked" over all by itself. No one has touched it to put it back in the pile.

I simply must add Solomon Islands to my bucket list.

He and Peter go out for a surf and the girls dip into the water, while Michelle and I take the babies up to the showers to rinse them off. A strange man tries flirting with Michelle... she's polite, but it's obvious she's married with babies. He runs off just as suddenly as he'd arrived and when we reach the towels, I realize my purse is missing. Was the guy distracting us so his pal could rob me? Ugh. Probably so. Oh well, the purse was a $2 bag from Malaysia and I only had $5 cash, plus my $2 fake student ID in it. I'm mostly bummed about the ID. When I decide to keep traveling, it would've saved me big bucks.

Tonight, at Brody's there's a fashion show, with a proper runway stage set up on the dance floor. It's a crazy, fun night because all the models are not wearing *anything* but full body paint. Even the men are fully painted but wearing tiny pouches. All I can see are nipples, cracks, and cheeks everywhere. *There must be some public law against this in the USA?*

Some of the staff is off work but are here partying. Standing right in my waitress 'well,' drinking right where I need to get my drinks from the bartender. *Rude.*

Elliott, that new manager who'd replaced Todd, has been so nice and has always asked if he can help me. So tonight, I ask if he wouldn't mind having the other workers step to the side so I can work. I gently tell him it feels awkward coming from me. I like everyone but haven't made any real friendships except Kate and my roommate Mason.

"Right away Pam!" he says, running off to tell them to move. Wow, what a difference in a manager getting the job done. I feel like he has my back. I balance my tray of glasses through the dancing drunken crowd, back to the bar to place more orders.

Everyone's still in my way. Only now, Elliott's joined in and he's doing shots with them. *Seriously?* I tap Elliott on the shoulder. He looks at me, his face reddens. I see five upside-down shot glasses in front of him on the bar.

"Can I please order some drinks here?" I ask with a small, annoyed smile.

"Come with me, Pam," he marches me to his brand-new back office.

I catch Mason's eye and shake my head. This is not going so great.

Elliott closes the door. "I'm going to have to let you go. This isn't working out." He's slurring his words. *Where's Brody when I need him?*

"Really?" I can't even hide my disgust. "You can't fire me. *Slaves have to be sold!*" I turn on my heel and slam the door on my way out. *What is with everyone?* I get home and realize I've accidentally stolen some poor dude's $10. Sorry, mate... got fired before I could bring back your change.

At home, Heater's still awake and Dwayne's just getting home from his shift from the cruise. They're bummed out for me, but together, we realize I only have a few weeks left until my ticket to Thailand. If only Jeremy had a phone, I could have visited tomorrow. Oh well, it's his turn to come up to visit me. I'd told him I can't afford $40 round trip bus tickets anymore.

Mason tells me a letter has come for me from a dear old friend from San Diego. His name is Sumit. He writes he'll be in Thailand if I want to meet up. He has no idea I'm already booked for that time period. I quickly pop a letter back with my flight details, hoping it gets to him on time. I won't have any way to track him down unless he's able to meet me at the airport.

"It reminds me of that book, the coincidences, the timing... it's amazing," Heater mentions.

She's right. I guess I'll just go with the flow. We hang out with our pals for the next few days, picking frangipani flowers, sleeping in, heading over to the candle-lit nighttime flea market.

After his visit, I follow Jeremy back to his house. We're just enjoying the days. The ram in the yard, drumming, meditating, painting, swimming, hiking, he even takes me to his Capoeria classes. It's a martial art from Brazil that looks like street dancing, but it's really fighting. I learn that *half of all slaves* brought over from Africa were sold in Brazil, not just America. When the slaves were shackled together, they had to learn to fight while they were hunched down, chained together, hands cuffed. There's a lot of balance and kicking in the face.

Jeremy's still teaching me all about his spiritual beliefs, and this feels much more right to me. It ties in with all of my travels and the things I've learned in life. *But… I get it.* I don't *always* want to have such heavy talks. I suppose it has suited me to be a hippie for three days a week, then head back to normal life. I'm a little *bored.*

Jeremy and I are at the market, and of course, I'm paying for the food while I stay. Usually, I'm make-up free at his house, with camping hair and a smile. But today, I decide to put on mascara and red lipstick, along with a deep red sarong with cream-colored peacocks on it, a white tank top and a wide-brimmed straw hat. My hair is sun bleached golden, even blond on the ends, and down to my waist at this point. I haven't had a haircut in two and a half years. I would take better care of myself at home, and I've been here long enough to not just "go camping."

A good-looking gentleman in a suit passes me in the aisle, looks me over, then does a double take at me as he slowly rounds the corner.

Jeremy sees this unfold. "I hadn't really looked at you in such a way, but to see that guy staring at you in awe, I see you in a new light Miss Pam. You look beautiful today, even with all the makeup on," he tells me, squeezing me into a hug. More of a claiming hug than a genuine affection hug.

"I'd love to come visit you in Los Angels," Jeremy tells me while we're walking home from the market.

"Los Angel-*es*," I smile. "I'd love to show you around. You'd love my family and friends! We could go to Disneyland, Venice Beach, Hollywood, even Las Vegas and Mexico are short drives," I get a quick vision of us getting married, having babies.

"Pam," he says, looking away, "I would *hate* all of those places. Couldn't you take me to that Yosemite? And don't you have, like, a desert nearby? I've heard of Yellowstone, and the Grand Canyon too..."

"Sure, of course..." It hits me then. *This is not the guy for me.* I love Yosemite and nature, but I'm also part Vegas, baby. And men in suits.

I think about how would he really fit in with my family?

I truly *look* at him. He's tall, skinny, black hair pretty much shaped like a mullet, and those bright green eyes. He's also the palest white skin color with zero muscles. He's dirt poor. And not much drive. And he's just so *serious.* He says he likes me for my sense of humor, like when he bought us both a Reese's peanut butter cups to share, but I'd stared at him, smiled, and ate his cup as well. He'd laughed and laughed. *But he's not funny.* I need funny. I'm hilarious and raunchy. I need hilarious and raunchy too.

We get back to his place and I realize I've been adopting his life every time I come visit. I miss *me.*

I sit on the front step to watch the roaming ram while he paints in the studio. I grab his boom box, take out his tribal chant music and put in, yes, you guessed it... *Metallica.* Not the new crappy stuff, the old, perfectly heavy '*Master of Puppets*' album.

He comes out, clearly ruffled feathers. He can't paint because the energy of the music is disturbing the very molecules in the oxygen. I explain I'm not in a bad mood or anything, I just miss my own lifestyle. He goes to hug me and a spider falls out of his hair onto my shoulder. Fear grips me as I scream and shove him away. He tells me they're perfectly natural and they're in his studio sometimes. He's just tall enough that he has to duck a bit through the doorway. As he's explaining it, a SECOND one climbs onto his collarbone. *GROSS!*

"You just made that happen with your energy, Pam," he tells me. "You're a powerful woman. Your mind is powerful."

I think he knows we aren't connecting this trip. I've come for a visit every week for months, but he's come up only three times. And now I'm not working. He gets it and offers to come up to visit Surfer's next time. Yay, my house suddenly seems *cleaner*.

I'm realizing my trip is coming to a close in Australia. This has been the pattern. No job, no boyfriend, a bit restless. That's my cue to move on.

*****

When Jeremy arrives on the bus, I walk to pick him up. We spend the afternoon on the beach with Davina, Michelle, Peter, baby Liam, and sweet Jade. Jade has been my little buddy. I've been carrying around packets of honey, so each time I see her I squirt a bit of honey onto her finger for her to pop into her mouth. She loves her Auntie Pam. She calls a Ham sandwich a 'Pam' sandwich. Peter had even said "Ha-ha-ha-Haaaaam sandwich," and she copied with "Ha-ha-ha-Paaaaam sandwich!"

They want lunch at McDonald's—Macca's—on the corner of the boardwalk. We all order burgers and fries, along with sodas and shakes. Jeremy, instead of ordering, unwraps plump, fresh, organic dates from his satchel. We all chit-chat about our shared adventures while I watch Michelle feeding Liam fry after fry. He giggles, reaching for more.

"He's just starting to like to try salty things... he loves new food adventures!" Michelle gushes about her perfect Gerber baby.

"Let's see if he likes dates," Jeremy says as he breaks off a piece. It looks weird, like a large brown raisin, sticky and chewy. He reaches over to place it in Liam's mouth. Michelle swats his hand away and says maybe another time. Jeremy just shrugs and eats his date.

Suddenly, in my own mind, I'm *all* women since caveman days. Each mother throughout thousands of generations feeding their babies first breast milk, then things found nearby like mangoes, bananas and papayas plucked straight from the trees, then snap peas, berries, taro root. Eggs and fish. And yes, dates as a sweet treat. Nowhere, in all the history of motherhood, has any woman until our last generation fed their toddler baby such rubbish as a McDonald's burger and deep-fried-in-lard-and-grease-nutritionless French fries. To swat away a fresh organic sweet date and instead opt for fast food for your baby, who is too young to even speak, is a real eye-opener. What are we doing to the children where *this* is normal and fresh dates are *weird*?

A couple of days later, I walk Jeremy to the bus depot. We've had a nice time, but maybe I'm pulling away a bit to prepare myself to leave again. My plane ticket to Bangkok is around the corner. Jeremy gives me a hug, tells me what a great time he had.

"I'll see you soon?" he asks.

"Of course!" I tell him.

He smiles, walks over to the benches, and sits down.

"Isn't your bus leaving soon? Shouldn't you buy a ticket?" I wonder.

"Yes, there's a bus leaving in five minutes and every hour after that until midnight. I don't have money for a ticket..." he tells me, still smiling.

What. The. Hell.

"What do you mean? How are you going to get home?" frustration evident in my voice.

"I'll put out positive energy and see what happens. If it's meant to be, it will happen," he tells me.

"Oh, okay, best of luck, darling. See you soon, bye!" *IS WHAT I SHOULD SAY.*

"GOD DAMMIT, JEREMY!" is what I *really* say. I slam $20 onto the counter, my face red with anger, full frown, feeling taken advantage of. "I'm ***traveling***!

I ***always*** come down to visit you, it was ***your*** turn just so I could save my money! *Really?*"

"Oh, thank you..." he seems a bit unsure of what to do next. *But he doesn't hesitate to take my money*, I definitely notice that.

I stomp away without looking back. I see him in a new light. I feel manipulated. If he really believed he could've made it home on *positive energy* alone, he should've seen my energy was ALL NEGATIVE, tainting the money. He shouldn't have accepted it. How come I was buying food for him on his visits here, plus pitching in a lot for the groceries? I even had to pay for my own Capoeira classes. And he didn't buy or paint me a Christmas gift after I was disappointed at not getting a birthday present. *You're welcome.*

*****

After weeks of beach days and, after making up, back and forth to see Jeremy, it's my last few days in Surfer's Paradise. I need a small haircut. Just a trim. My hair is down to my butt but it's pretty ratty. I haven't blown it dry the entire trip, except for that couple of days in Queenstown. One of my customers from Brody's is a hairdresser, so I pop in to see her for a trim.

"So, about an inch, I love the length, and the V-shape," I tell her.

"Yep, mm-hmmm," she says, getting her scissors ready.

I relax into the chair, magazine in my lap, and loving that it costs just $6, yippie.

She cuts a big chunk AT MY SHOULDER. My jaw drops open.

"*I just wanted a trim... remember?*" I say faintly.

"You didn't want all that blonde on there, that's just damage," she tells me firmly, snapping her gum. I can't look away from the floor. 15 inches of my gorgeous golden-blonde-traveled-around-the-world-surf-beach-adventure-sun-kissed-cinnamon-nutmeg-flowing-youth is on the ground.

I feel the sting. I ***earned*** that hair. I ***earned*** that blonde.

I place the money, *plus a tip*, on the counter and whisper a "Thank you" as I'm running out. Here come the waterworks. The tears begin to flow.

I hate it. I hate her. Now it curls up even shorter without the weight. I look like I have shoulder length curly poodle hair. I'm so upset. There's nothing I can do. I don't want to ruin the last few days here. So, when all my friends tell me how cute I look, I plaster a fake smile on and pretend I'm Rapunzel, not Shirley Temple. Sorry locks, see you in another three years. I vow to continue to travel and not cut my hair forever.

*****

My last day in Australia. My backpack's full of clean clothes. Everyone stops by to see me off. Hugs to my friends who made me feel like an Aussie. Finally, I turn to Jeremy. He came back for a few days to see me off. We kiss for a moment and just hold each other. He has changed me. I'm a better person for knowing him. Falling in love overseas feels like a romance novel.

"I brought you something as a going away gift..." he tells me smiling. He pulls out an aqua green suede pouch and opens it up, dumping the contents into my hands. "There are three porcupine quills from my mother's farm, so you remember our time together there. There's a piece of turquoise from one of my art shows to bring your soul peace and calmness. There's a white stone from the river on our hike that day," he giggles. "This last one is a piece of solid amber. It's a stone that brings you good fortune. And I wrote you a love poem, to remember me by..." My heart swells. This was all I'd wanted.

"Thank you so very much! I'll treasure this time with you. If the world allows it, our paths will cross again very soon," I say, my eyes welling up with tears as we hug for the last time. He whispers something in my ear as my taxi pulls up. I smile, even though I don't catch it.

I'm mentally in Asia already.

Thailand

Bangkok

- Peace Train -
Cat Stevens

Cat Stevens flies me to Thailand. As I land, I notice the city is *massive.* It's like landing in Los Angeles. You fly over it forever. My last letter to Sumit, with my flight number, was weeks ago. I hope he's at the airport or we are never going to find each other. As I pass through customs, sweating bullets in the humidity, still with my tiny school backpack, I hear him.

"Pam!" he shouts, waving and smiling.

Sumit was a regular customer at that fancy restaurant where I worked in San Diego, California, a few years ago. Yes, the one where my truck was stolen. We'd stayed friends and he loved that I was traveling. He's a handsome man of India-Indian descent, but raised in America, so a Yankee accent. He has straight black hair, brown eyes, and a square jawline. Just a touch taller than I am, and always with a big smile. I love that he's joining me on this adventure.

I'm so excited to see him, and I run into a hug.

"This is all you brought?" He furrows his brow. "I can't believe it... not bad for a girl!" He pokes me in the ribs.

It turns out he arrived earlier today and has a hotel already for us to share. Then tomorrow, we can take the 14-hour train ride up to Chaing Mai and Chaing Rai, near the Burmese border. Sounds good to me.

We head out into the sticky night air. Smells of fried crickets, sweet mangoes, fish, and raw sewage fill my nose. I love that it stinks. It so reminds me that I'm in an exotic land, far from everything I know. Australia was across the ocean from my home, but cities along the beach? Blond people everywhere? It reminded me of California.

Sumit hails a rickety tuk-tuk, and off we go, swerving around buses and pedestrians. I notice the 10th vehicle we pass. There's a cat on a ladies' shoulders on the back seat of a motorcycle—oh—and she's sitting side-saddle, reading a book. *As you do.* The 14th motorcycle we pass is piled 20 feet high with hay.

The next interesting thing is *us.* We veer into the *oncoming* traffic to get around the stopped traffic in front of us. Our driver decides to push his way through, making them veer out of our way. *He seems to know what he's doing, right?* I guess we're in a rush to get to the hotel, death be damned.

The room is much nicer quality than I'd been used to. Marble floors, two queen beds with mosquito nets, a balcony. And dare I say it? *Air conditioning.* And, yes, of course, the fan favorite, the free banana pancakes. Oh, I've missed you, Asia. *With rice... noooo!*

Sumit yawns. "I'm still so jet lagged. I need some serious sleep to enjoy myself." We get some sleep and I wake up in the night to throw up. I wasn't even feeling sick. Hmmmm, must've been the plane food or that tuk-tuk ride or something. I don't mention it in the morning, because, well... *gross.*

"I want to hear all about Africa," Sumit says, biting into his pancakes.

"Do you want to hear the story of me almost getting stabbed on the train my first night there? Or that I learned to surf properly? Or how Spanish saved me from being gang raped? Or how I was chased by a herd of wild elephants? Or even getting high and wandering around a wild animal park, forgetting that it was a Wild Animal Park? Or how I sold T-shirts on the beach during *The Gunston 500*?" I sip my tea, eyes twinkling.

"Oh, yeah, all of that. It's definitely on my bucket list!" he says, clearly proud of me.

After we finish up, we walk down the street a bit, checking out the vendors. Every square inch of sidewalk is covered by flea market booths. Prada purses for $4 *(Yep, real thing guaranteed)* Armani jeans $3 *(Quality, baby)* with scarves, wraps, jewelry. I even spot the old familiar watermelon-juice-in-a-plastic-baggie with a straw. I buy a bottle of water for the train ride. I forgot how to haggle. I pay quadruple the price. *Just so thirsty.*

We buy our overnight train tickets and settle into our cramped seats. *Yay, 14 hours.* Well, I haven't missed *all* of Asia. Moments into the ride, a conductor comes up, asks us to collect our things and follow him. I look at Sumit, and he shrugs while grabbing his bag. He leads us to first class, bowing with his hands together, wearing an enthusiastic grin. He insists we take the wide, cushy seats so we're more comfortable. *Ummm, okay. Is this allowed?* That was kind of him.

I secretly worry about those rumors you hear of western travelers getting mugged for their backpacks on the overnight train in Asia. *Ah, well, this seat is comfy. I'm sure we'll be fine.*

Six hours in, I feel my cheeks sour. No tummy troubles, no nausea, and yet, here comes lunch. I move lightning fast to grab the plastic bag from my bottled water purchase. I barely make it in time and heave into my bag, over and over. Great... somehow the conductor poisoned me. The other Thai passengers look away. *Sorry. I'm just a human.*

I glance over at Sumit. Awesome, he's asleep, head against the window, slightly snoring. I'm unsure what to do. I've got my bag of barf in one hand, and our backpacks with everything we own in the bin overhead. This train is packed full, I feel uncomfortable leaving the bags while Sumit's sleeping. Plus, what can he do? I just sit there with my barf bag, scrunched tightly. Talk about *gross!*

About an hour later, same souring cheeks. I just open my only bag and go to town. I'm so happy that I didn't dispose of it after the first time. *Pat on the back for my laziness, yay.* About an hour later, I go again. I'm single handedly pulling off the blueberry pie eating contest from '*Stand by Me.*' My bag is getting full... and heavy.

Finally, Sumit wakes as we arrive. I have no choice but to tell him.

He looks at me, "Oh no, what's wrong?"

"Actually, I'm feeling fine, totally fine... but..." I hold up my barf bag. It looks like a loaf of bread. "I've thrown up eight times since I landed!" I'm mortified.

He goes on to ask the array of questions best left between ER doctor and patient.

I swear I feel fine, I just can't keep things down.

Chaing Mai

- To Be a Star -
Cat Stevens

We grab a tuk-tuk to our hotel. The driver won't stop smiling at us and bowing his head. Next to us, a man's *riding an elephant with a red painted face* down the highway. Arriving at the hotel, the driver offers to carry our bags in. That hasn't happened to me before, maybe a Thai tradition?

The hotel is all dark wood, batik paintings, carved dancing Thai statues. A large patio area with couches and fire pits, strolling paths and koi ponds. *Ahhhh.* The check-in clerk is all smiley and bowing to us... he keeps shaking Sumit's hand. Our room is nice again, white marble floors, and delicately carved dark wood furniture. I heard Sumit say the cheapest room. This is the cheapest room?

I splash some cold water on my face and take a chug of water from my bottle. I feel normal. Not even a tummy flicker. I think back to the last time I ate. I'm probably just hungry. Oh no. I think back to being with Jeremy... *I'm not... I couldn't be...* we were responsible every single time. I can't worry about that right now. *That's Future Pam's problem.* I must have caught some bug from the airplane. Within seconds, my water comes back up. This is not good.

Sumit decides to go get us some food, something light for me, like dry white rice... *dammit.* It too doesn't stay put. This goes on through the next morning. Even just a sip of water and two seconds later, I'm like a water sprinkler on a golf course.

I decide it's mind over matter. I'll try to walk it off and go with Sumit on an excursion. We climb into the back of a large military vehicle with a handful of other tourists to go see a snake farm. About three miles down the road, the movement of the truck, the fumes, the heat... *I feel it again.* I climb over two people and just make it to the edge of the open canvas to hurl green slime onto the windshield of the tuk-tuk behind us.

"You don't look so good..." Sumit tells me.

"I thought you might be sick..." a woman tells me in a thick French accent.

"You are the color of green," a man from Denmark tells me.

Another man knocks on the back of the driver's window, tells him to pull over. Here we are on the side of an unbelievably busy highway, I feel embarrassed and defeated as I climb out.

"I can't go," I say, humbled. I guess I'm sick.

"Well, of course I'll be coming with you," Sumit says, climbing out as well.

I hold up my hands. "No way, José. I absolutely insist you keep going. Seriously, Sumit, I'm going back to sleep off whatever this is. What are you going to do, watch me sleep? I won't let you miss out. You only have a few weeks before you get back to work. I'm continuing on with my world tour." I smile and nod to show him I mean it.

Three tuk-tuks and five motorcycles pull over and gather around us, excited to see some action. We agree to split up for a few days as he goes on some adventures. I hop on the back of a mysterious motorcycle whose driver says he knows how to get me back to my hotel. *Giddy-up.*

The next few days are a blur. I sleep, and sleep, and sleep some more. A sip of water still won't stay down, but I have to try. On day three, there's a timid knock at the door.

I creak, "Yes?"

A red-head lady peeks in. "I heard you're sick," she says in a thick Irish accent. I nod.

"I think you have what I had..." She digs through her satchel and hands me a packet of pills. "Here, take one every four hours, you'll be okay!" she smiles warmly and closes the door.

I stuff one in my face and pray it stays down. I just chew it and swallow without water because I can't afford to waste a single pill. I didn't even catch her name. I get a funky thought about taking drugs from strangers. *Sorry, Mom.* I promptly pass out asleep again.

Day four and I already feel better. I do a big stretch before climbing out of bed. There's a note under the door from the clerk that says Sumit will be back this afternoon. I feel so absolutely stiff from all this lying around. I'm ready to tackle the day, so I decide to go get a massage.

Sunlight pierces my eyes. I walk a block and find a Thai massage place. I can't believe the prices. $2 for an hour or $3 for two. I go for two hours. The young Thai girl, maybe she's 17, brings me to a room and shows me some silk M.C. Hammer pants to put on, no shirt, no towel. She doesn't leave while I undress. We smile at each other while I step out of my shorts. *So sorry about my hairy legs, I've been sick.*

She puts me face down on a mat on the wood floor. She starts by pulling my toes, pretty soon she's folding my legs back to my butt, smacking my kneecaps, after that walking on my neck, karate chopping my glutes, kneading my boobs, even pulling my hair. All of it sounds weird and excruciating, but I love every minute of it. She's got my legs up in the air, over my own ears, as she pretzels me. I buy a third hour. I so needed this.

After we're done, she watches me change back into my clothes. We smile, we nod. I hand her $3 extra. She shakes her head no, looking confused. She tries to motion that I already paid. I put my finger to my lips like *'it's our secret'* curling her fingers around the money, nodding yes. Her eyes well up with tears and she does a deep, meaningful bow. I feel like a million bucks.

I just hope it wasn't a sex place and that she knows I *really* did only want the massage. *No sucky, no fucky, no love me long time.*

I head back to nap a bit more after my kick-ass massage, but suddenly, I'm STARVING. I could eat my own arm. I turn into a side shop with meat carcasses hanging in the window, flies flying around. *Lovely.* I walk out. *I can't.* I buy a few lychees to start small, just in case.

The military truck arrives back with the gang that I was supposed to get to know. Last, but not least, I see Sumit. While everyone else is in beige baggy shorts and T-shirts, he's wearing neon shorts, no shirt, neon basketball tennis shoes, and mirrored cop sunglasses with a Crocodile Dundee hat. He looks like a rock star from the 80's, like he's making an entrance. I realize he's having a moment. He's feeling *sexy.*

He tells me he rode an elephant. While his friends' elephants strode through the river, his tiptoed along a fallen log, like an old woman who doesn't want to get her hair wet.

He's so happy I'm feeling better that we decide to go out to a fancy dinner and a show. We dress up, I put some makeup on, and break out one of my Bali sundresses. He cleans up nicely in a white button-down collar shirt and jeans. We walk to a fancy restaurant, overlooking the river. There are intricate carved statues and paper lanterns, along with dragon icons everywhere.

It's a little crowded so we must wait for a table. A man rushes over, introducing himself as the restaurant owner, shakes Sumit's hand, and bows deeply. He shouts out some things to his staff and they quickly stop everything they were doing to place a new table and chairs right in the front of the stage. We get escorted through the restaurant like Ray Liotta on his first proper date with Karen in '*Good Fellas.*'

"All right Sumit, what's going on? I've been all over the world and have never been treated like this!" I inquire, placing the cloth napkin in my lap.

He grins. He laughs. He can't help it.

"I know, right? You know how in America we think Hollywood is the movie making capital of the world? Well, it's actually Bollywood, in India. They produce more movies and distribute to more places around the globe. They think I'm a famous Indian guy. A genuine top Bollywood actor. I've been told I look like him my entire life. I'm also dressed in American clothing, I'm traveling overseas, I'm with a white woman... we keep getting the Bollywood movie star treatment!" His eyes twinkle as he chuckles.

"Oh my God! That's hilarious!" I cover my mouth to hide a loud laugh, I can't believe it. It all makes sense now. Everyone has wanted to shake hands with just him, not me. They bow to him, they come running over to assist, they upgrade the things he's been buying.

Just as dinner is served, stunning Thai dancers wearing brightly colored dresses, dripping with golden jewelry take center stage. The plinky-plunky music begins as they do the bending of the fingers, the tippy-toe dancing while slowly turning and bending. They smile and giggle at us in the middle of the dance.

I can see the manager hovering just off stage, give the dancers a Look. He's signaled the girls to come down off the stage and dance around Sumit and me.

We giggle, they grab his hand and lead him on stage and make him dance with them.

I can't stop laughing... the crowd thinks they recognize him and chant and cheer. After Sumit swivels his hips a few times and tries wiggling his fingers to no avail, he joins me again at the table. A small crowd gathers around him. They want to touch him, meet him, some even just stare at him.

We lock eyes, stifling laughter. What a scamp he is for going along with it!

*****

Chaing Rai

- Macarena -
Los Del Rio

We're back on the military truck with a few other tourists, driving through stunning rice paddy fields, pineapple farms, and thick jungle. We're heading north past Chaing Rai, to the hillside tribal village well known for the mysterious Kayan people. Upon arrival, we're greeted by a local guide, who translates for us.

The Kayan are refugees from Burma who were originally granted a "stay" to avoid the war, but they've now become a famous tourist attraction known as The Long Neck Women of the Hill Tribes. They wear the brass rings around their necks, beginning at age five, adding a new ring each year until their 21st birthday. They feel the longer a woman's neck, the more beautiful she is.

Strolling through the tiny town, we're observing, smiling, nodding. Women sit along their front porches of their huts—tiny homes made of sticks that could blow away in a small storm—weaving palm fronds into baskets, selling beaded jewelry, carving souvenirs. The women are dressed in beautiful colors, head-wraps, with those brass rings around their necks. The older ladies' necks with 16 rings, the younger girls starting out with just a few. I'm so happy to meet these girls, ladies, women, mamas, grandmas.

They're open to answering questions about their tribal customs. I'm torn. It feels like we're exploiting their differences, and yet, they need the money from the tourists or they don't eat.

We learn the brass rings are quite heavy. The girls say it took a while to get used to it, and even now, they still get bruises on their collarbones. They sleep with them on, eat with them on, bathe with them on. It seems so... *itchy*. Like the dirt and sweat from their day is trapped under the rings for 40 years.

And are they really into this custom, to meeting people from all over the world? Or are they being forced to act like zoo animals and perform for the tourists, day in and day out? *What do they think of my neck? Is it short, squatty... even hideous?*

We talk to the women for ages, offer to answer questions about where we're from. We discover one family has 13 children and would we like to **purchase one child?** *Wow... no thank you, maybe next time we come through town.*

Sumit questions our translator, "Having 13 children must be very tough, very expensive. Do they want to sell a child to be able to buy food to feed the other 12 kids?"

"No," he says. "They want to buy a Jeep. If they can sell one or two kids to foreigners, they can get one." He smiles, nodding in a way, as if we understand that dilemma. *Yep, we know how it is. When you want, like, a TV or a new couch or something, just have another kid. Preferably the kind you don't mind selling to strangers to do whatever they wish to your precious baby.*

We learn they had three other children that have been sold already.

How absolutely tragic to think where they must be.

My heart skips a beat as it physically breaks for these missing children.

A dark cloud hovers over the conversation.

The guide says the Village Chief would like to meet us.

Why not? When in Rome?

We meander into his hut, he's squatting in that Indonesian style and yes, his nards are hanging out. He lights up, takes a puff, and hands us the long wooden pipe. The guide says it's a tradition to smoke opium here. *Do I really want to be an opium smoker? No!* I just hold a tiny puff of smoke in my mouth and pass back. He's not saying a word, so we aren't either. We smoke 'em peace pipe for a few puffs and pass it back. Does he just do this all day, every day? *He's a real giver, it's for the tourists after all...*

He finally mutters something, and our translator turns to Sumit.

“The Chief is asking you for your autograph, he knows you are a famous actor in the Bollywood movies...” he smiles.

Sumit signs his name along with a short note about enjoying the company of the great people of the village. I bite my lip to not snicker.

“Do you feel anything?” I ask Sumit out in the sunshine.

“No, you?” he replies back.

“Nope.”

“Must be some flimsy weak bullshit opium for the tourists. Probably smoked some kind of tea or something!” We burst out laughing, but genuinely, not high in any way. We got ripped off on our opium purchase. I guess the chief knows we’re not exactly going to the cops with this one.

“And great job,” I tell Sumit, “You just gave your movie star a drug reputation with ‘street cred’ in the high hills of Chaing Rai. I hope it doesn’t go world-wide and end up in the tabloids!”

With a few of our new tourist friends, we show up to the Full Moon Party at an outdoor patio bar, complete with a neon black light, a disco ball, and some techno music with laser beams matching the music beats. It’s connected with what otherwise would be a Gilligan’s Island lunch-hut... but on a night like this, they’re ready to party. Locals and tourists alike start drinking, dancing. Everyone of course wants to hang out with Sumit. They take photos with him, buy him drinks.

I hold back a bit. Yes, I’m with a trusted friend, but I’ve heard so many rumors about the aftermath of Full Moon Parties. The bad guys know you’re wasted, and it’s the perfect time to rob you of your backpack, your money, your kidneys, even **you,** to sell on the black market. I could be kidnapped and sold as a sex slave. It is Thailand after all. I have a beer.

We’re dancing with new friends when the ‘*Macarena*’ comes on and we all do the steps properly, no matter which country we’re from. With the scent of the jungle, a throbbing strobe light, in the luminous full moon overhead, a tribe of small monkeys runs through the bar.

It is stunning to be on this exotic adventure.

Bangkok ...song

- Planet Caravan -
Black Sabbath

Back in Bangkok, over the next few days, Sumit wants to do some shopping and sightseeing. We hire a couple of motorcycle drivers to take us around for the day. We head off across the city to the famous Amphawa Floating Market. Weaving through the crowded city streets, inhaling bus fumes, I can taste the thick grimy air, and I'm holding on for dear life. After what feels like hours on the back of the sweaty, stinky, loud motorbikes, we see glimpses of river life.

It looks like the '*Waterworld*' movie set. Hundreds of boats barely squeezing by each other. Each craft filled with colorful wares. Clothing, hats, cold sodas, fruit, veggies, fish, snacks, carved wooden knick-knacks, religious deities, even music CDs. Everyone wears cone shaped hats to protect them from the sun. I can see the warmth of their smiles. Quaint little homes line the river's edge, near a tiny ancient temple nestled into the jungle.

Sumit and I hire a boat ourselves while our taxi motorcycles wait for us. We head out into the open market and marvel at the fact that these folks have been doing this river life for 1000 years. The heat, the cooking food, the swampy smell of the river, the overall hum of everyone speaking as they do business. A child sleeping in a hammock nearby. Squeals of laughter, dogs barking, and a thousand motorcycles in the distance. We just take it all in.

We purchase a few bananas, a couple of sweet treats, a few carved trinkets. I think Sumit gets stuck with the movie star price, and, for the first time, it's not so much fun.

Our next stop is the Wat Pho Temple, a Thailand museum of history and culture. I cover my shoulders as we enter the temple, and it does not disappoint. There are massive buildings inside the gates, all with golden rooftops and intricately painted designs. Large painted cone-like creations hold the ashes of the royal family. The entire place is home to more than a thousand Buddha images. Hundreds of golden deity statues, all with hands pointing up, down, folded, touching the ground, facing his palms upwards... each means a different connection with the universe. The most sacred building in the complex houses the three-tiered gold and marble pedestal that is still used in religious rituals. Today, there are 20 monks meditating.

Then, on to the Reclining Buddha. Not only is this famous Buddha statue massive at 150 feet long, but the intricate carvings of his spiky hair and the mosaic designs on the bottom of his feet make it surreal. The meaning of this pose is that he has reached Nirvana and will no longer need to reincarnate. The statue has a brick core, and it has been covered in gold. The bottoms of his feet are encrusted with Mother-of-Pearl, 108 inlaid panels of auspicious symbols portrayed at various stages of love and light: flowers, dancers, white tigers, elephants. The Buddha itself was built first, because they didn't know what the final size would be, so the temple building was framed around it. Completed to exquisite detail, it's known as the First-Class Royal Temple.

The Buddha, of course, is famous for reaching enlightenment while fasting under a Bodhi tree after leaving his sheltered, wealthy life behind and viewing poverty and suffering for the first time. I feel honored to leave flowers for his memory near the burning incense.

It feels odd to step out of the peaceful vibe of the temple and smack dab into the roar of the city once again.

On our last day together, Sumit and I spend the day scooping up rare finds to ship home. We tuk-tuk to the Pat Pong market in the center of the city. It's like a giant flea market. I purchase a bronzed metal three-foot-tall candelabra with matching leafy mirror, a huge hand carved teak wooden elephant, and a couple of two-foot-tall giraffes. I even pick up a green wooden Egyptian Pharaoh to someday put in my garden. I get shirts, belts, jeans, purses, and things for my family once I return. Sumit gets an Asian Gong, and a four-foot-long intricately carved black dragon. Back home he's going to slap a slab of glass on top so it can be his coffee table.

"I'm going to feel like Indiana Jones when I crack open the lid of the shipping crate when it arrives!" he tells me, giddy as a schoolboy.

"I'm already jealous that you get to be there to do it," I laugh.

It's going to take a couple of months to arrive into San Diego's port and will have to pass through Customs, being cross referenced with import and export laws. Who knows where I'll be?

As the sun sets, we decide to go to a Muay Thai boxing event. Of course, while we're trying to buy the tickets, they refuse to let us purchase the regular priced ones and insist on giving us the VIP tickets. VIP at a Muay Thai boxing match is a chained-off section next to the regular seating. Same broken, hard, splintered, wooden bleachers. No Smoking signs everywhere, probably so the boxers can give their finest performance, yet 99% of the people have cigarettes dangling from their lips. Everyone's passing cash back and forth, yelling their bets on who'll win. I'm the only woman in a sea of 300 men.

The fighters are introduced in the ring, already sweaty, because there's zero air conditioning. They look like solid welter weights, but as soon as the bell rings - *WHAM!* An amazing kick to the shins that sweeps the opponent down... he's up, there's an ear jab retaliation.

I love it. It seems like American boxing but with no rules. Wild. *Animal Style.* I've never seen anything like it, and I'm instantly hooked. The boxers are ripped and strong. They go for five rounds before Orange Shorts is announced the winner. He's the underdog, so the crowd goes crazy.

We find ourselves shuffled along to hop into the ring to meet the boxers. Everyone wants a picture with movie star Sumit, and me too, this time. I guess they don't get many tourists in here, especially women. Men are swarming us as we walk through the crowd. They're grabbing at my face, my neck, my hair, my waist, yet not in a threatening way, they seem to be posing for boxing photos.

We're all giggles with these extremely drunk men.

Did they gamble all their family grocery money on this?

Do their wives know they're here?

*****

We eat a gorgeous breakfast, relaxing in the sun just before Sumit's taxi arrives. We share the biggest hug. I tear up, like when Jay left me in Africa, and Sandy in Australia.

Traveling brings out the realness in people.

After Sumit drives off, I walk into the nearest travel agency office and grab a flight to Los Angeles, with two major stopovers in Bali and Hawaii. I'm heading home. *Sort of.*

First a few days of massages and banana pancakes.

I'm sure going to miss traveling like a movie star.

Indonesia ... dua

Bali ...tiga

Kuta Beach ...tiga

- Strangers -
Portishead

Ahhh, Bali. It really does add a comfort to know where to go and what's expected of me. It's pouring with a warm, tropical rain. I go straight to Poppies Lane Two in Kuta Beach, to my old hotel. But this time, the road that leads up is fully flooded. The tuk-tuk driver drops me off half a mile downstream. I have no choice but to wade up to my knees through the muddy, mosquito-filled, mucky, stagnant, sitting water. *Isn't this how people get diseases and tapeworms? Disgusting.* But I don't have the time, desire, or energy to look elsewhere.

I even get my old room.

The last time I was here, I'd just found out my Dad had died. I think of the journey I've had since. I know in my heart I made the right decision to keep going. I truly felt my Dad had placed people in my path to help guide the choice. I thought I would be sad, but now I just smile. He'd be so proud.

I duck out of the rain, stepping into a shop called Bobocraft. It's filled with hand-carved teak drums, mandolins, and didgeridoos, and even small, colorful animal statues.

"Hi there," I hear a deep voice. I turn to see an older Indonesian man with dreadlocks down to his knees. "I'm Ketut. I own this store, please take your time," he smiles.

I love everything he's made. After Jeremy's drumming, I'd fallen in love with the thought of my own drum. They're all gorgeous in their own way, bright colored, jingly-jangly-dangly bits, sequins, and glitter. Then, out of the corner of my eye, from way in the back of the room, I feel one drawing me in. Like the '*Indiana Jones 3*' scene when he's searching for the Holy Grail and must choose among 50, he reaches for the simplest one, because Jesus was a carpenter. "*You chose wisely...*"

It's a D'Jembe, which is an African style, while all the Asian ones I've seen have skins on both ends with a wide bubble in the middle, like a barrel with small, covered ends. This is a dark teak wood, standing on a slim base with a much wider... Bowl? Bucket? Whatever. It's covered on just the top with a goat skin tied down by ropes around the edges. It has hand carvings of two dogs kissing on the edges. Here in Indo, a mainly Muslim country, I know "they" hate dogs here. Sadly, the stray dogs are covered in ticks and fleas, and there's lots of mistreatment. The island of Bali is a melting pot, but the main religion is Hinduism. It's rare to find anything about dogs at all. I ask about it.

"Yes, this is an African style, I used to live in Africa years ago and brought this type of design to my work. I was raised to think dogs were the devil, but, on my travels, I've seen so many other citizens of the world enjoying them as pets. I also wanted to bring that home with me. Maybe my art can heal some of the dog's souls that end up being treated badly?" He smiles again, offering me some mint tea.

I accept and end up buying the drum, a carrying case, a didgeridoo, and a few trinkets.

Ketut asks me to stay for lunch, telling me he has an American girlfriend and they share a child.

"Oh yes?" I'm intrigued.

"Yes, I tried to live in America, but it was in Michigan, and I just can't live with snow," he shivers. He points to the floor, just to the side of the front door. "She was 19 and I delivered the baby here in the store four years ago." I notice he seems to be in his 50's.

"Wait, what?" I gulped my drink. "*You* delivered? What happened? How did you know what to do?"

I picture an American teenager lying on the floor, delivering her first baby, away from everything and everyone she knew. *Wow, baby.*

"Oh, I saw it in a movie one time. I boiled water, got towels to make her comfortable. She pushed for hours, then I saw the head crowning and I grabbed it and helped pull her out. I had a piece of bamboo and I broke a slice off to cut the umbilical cord with it." He mimes slicing in the air, giggling like he's drunk. "It was the most beautiful moment of my life." He shows me a picture of a stunning blonde cheerleader type and their sweet, beautiful daughter.

We talk for ages and he brings me his favorite trinket in the store. Small enough to fit in my hand, it has ombre red, orange, and yellow fading together and it's light as a feather. I can't tell what it is. He tells me it's a hand-carved bird's beak. It's the lower beak of the Toucan, but it's so delicately carved it looks like lace, with the end curved into a curly-cue.

"I burn the end and gently nudge it, let it cool overnight, then do it again each day for weeks. I used to rush it, but it would break. Here, I want you to have this, as my gift. We are friends." He smiles his warm smile.

We hug goodbye and I take all my gear through the muddy river to my hotel and sleep for the rest of the afternoon. I thought I was brave, but a teenager had her first baby on the floor of a shop with bamboo slicing the cord.

That's Big Brass Balls Brave.

An afternoon of shopping brings me to a custom suede designer. The shop smells of leather and incense. I fall in love with a deep, jewel-toned, dusty maroon fabric. I choose the pattern of a long, sleek, silk-lined suede coat, with a fitted belt, and matching slide-on heels. They measure me for a custom fitting, trace my foot on a piece of notebook paper, and tell me to come back in three days.

When I return, to my utter joy and surprise, they've made me the perfect fit. For $40, instead of a backpacker, I look elegant for the first time in years. I belong at the Ritz in Monaco. It's too hot and fancy for here but will be perfect at home.

This is my third time in Bali, and I don't feel like sight-seeing. It's raining so much, I guess I'm in the rainy season now, I just end up just sitting in cafés watching Blockbuster American movies: '*Beetlejuice*,' '*Commando*,' and '*Weird Science*' to name a few. It's hilarious to see the Balinese translations. They don't have as many words in their vocabulary, so they miss half the jokes of the movies. They don't seem to mind. I don't meet any new groups and that's okay. Everyone seems wasted all day, and I'm not in that space anymore. Plus, my money's dwindling again.

A few weeks later, after just living my life Balinese style, thoroughly melting into my tropical paradise self on the Island of the Gods... I'm on my flight home to the USA.

United States of America

Hawaii

Oahu Island

Honolulu

- Numb -
Portishead

As the airplane wheels touch down on the runway, I see the American flag.

My chin quivers. A single tear slides silently down my cheek and drips onto my wrist.

The Brady Bunch, Gilligan's Island, I Dream of Jeannie... Sixteen Candles, Ferris Bueller's Day Off, Point Break, Star Wars... Wooly Bully, California Girls, Yankee Doodle Dandy, Shave and a Haircut: Two Bits... Bigfoot, Area 51, Roswell... Apple Pie, Del Taco, 31 Flavors... Malibu Barbie, Hot Wheels, Lincoln Logs... Bugs Bunny, Snoopy, Rudolph the Red Nosed Reindeer... Disneyland, Sea World, Magic Mountain... Academy Awards, Super Bowl, World Series... Hollywood, Venice Beach, Big Bear, Las Vegas, New York, Grand Canyon... Wonder Woman, Batman, Superman... Memorial Weekend, 4th of July, Thanksgiving...

Sigh. My chest aches as I put on my backpack again. *I feel patriotic.*

I change my Thai Bhat and Indonesian Rupiahs back into United States Dollars. The woman hands me $148.88. Yeesh, that's all I've got left. I just hold the money in my hand for a moment. It looks *fake*. They all have huge Presidential faces on them. This isn't even a good fake.

"Um, is this right?" I ask.

"The total is correct," she points out her math.

"No, the faces, they're huge...?" I reply.

"Oh," she stifles a laugh. "They changed it all, it's real."

"When?"

"Oh, maybe a year and a half ago."

Whoa. It feels surreal. I study the rest of the money. *Damn, dimes are Teeny Tiny. Have they always been this thin?*

I call a hostel in the North Shore and they send a shuttle to pick me up. It starts raining that tropical rain while I hop in and meet other tourists heading that way too. We pull onto the freeway, and as a surfer hears where I'm headed, he whispers in my ear, "I've just come from there, there's literally nothing to do there but surf. There are no markets, bars, restaurants, nada."

"Oh no," I whisper back, "That's not what I had in mind. What should I do?"

"I'm getting off at the next stop, in Waikiki Beach, at another hostel, just get out there too," he says.

The driver pipes in, "What's that, now? You're not staying with us in the North Shore?"

Gulp. I have no choice but to come clean. "I'm so sorry, I've never been before, and I thought there was more to do up there. I think I'd like to just get off with my friend here," I tell him.

He grips the wheel, swerves over to the shoulder, and stops the van. He jumps out, enraged, his face twisted to a scowl. He runs to the back and throws my backpack and new drum right onto the muddy side of the road.

"Get out! I'm not some taxi! I only pick up customers! GET OUT!" He shouts at the top of his lungs. His face is dark red, a vein popping out of his neck.

I climb out timidly, explaining, "Oh, I'm so sorry, I could pay you to ta--"

"FUCK YOU!" he slams the door and peels out, leaving me stranded on the side of the freeway in the Honolulu downpour.

Fresh off a 13-hour flight, I've set foot in the USA for exactly 10 minutes and this is how I'm treated? So much for the welcome mat, or getting lei'd...

I find myself excruciatingly exhausted, to the bone. I have no choice but to pick up my bags and stick out my thumb.

That moment, a dark green beat-up pickup truck pulls over and a little Vietnamese man jumps out, puts my bags inside the cab behind the seats while he agrees to take me to the Waikiki Hostel near Diamond Head beach.

I'm replenished. I feel my energy warming back up my body. Of all the people to help me in the USA, the first person is a friendly Asian. Is this a sign? Did I come back to the States too early? Too late now.

As I'm checking in at the open front desk area, the minivan pulls in that I got kicked out of just 20 minutes ago. The driver sees me, squinting his eyes in rage. *Fuck me? Looks like you could use a little Aloha!* I smile and wave. I beat you here fair and square. And I'll be sure to spread the word about how your hostel treats people.

My new surfer pal climbs out, laughing as the sourpuss drives off.

I check in and hand over my $140 to pay for seven nights. *Holy finances, Batman.* I have $8 for eight days. I need a job yesterday. The hostel just changed their policy from free breakfast to no breakfast. Am I going to have to buy a $2 bag of rice for the week?

The room's basic with 10 bunkbeds, so I'm sharing a room with co-ed strangers again. It's filled with very young, or immature shall I say, 20-year-old dudes. They're drunk, this is their awesome adventure, just being in Hawaii. I must remember that.

"Hey," a scraggly haired red-neck points to my drum. "What's in there?"

"It's a drum," I say and smile, hoping he goes away.

"Let's see it." He climbs off his bunk. Everyone gathers around me.

I slowly unzip it and display it for all to see. I don't really want to show them. I feel... *vulnerable.*

"What are these designs?" "Where'd you get it?" "Can I play?" "My turn!"

I humor them, tell them about the dogs, Bali, sure, okay.

There's a knock at the door and the front desk lady comes in to tell me there's now a bed available in the next room over, and it's all girls if I want to take it. *Yes, please!*

I start to pick up my stuff and Original Redneck puts his hand on the drum, telling me, "We'll just keep it in here with us. To make sure it's safe."

I stare at him, not smiling, and grab it, zip it, and walk out to the next room. If she hadn't knocked, was I getting robbed? Now I'm nervous to leave any of my stuff. It's been an unwritten law not to mess with people's stuff. I decide to let go of my discomfort and fear of getting anything stolen. I had everything stolen once upon a time and I'm still okay.

There's a Burger King *(yuck)* in walking distance where I guess I can get an 89-cent child's burger. Technically I can get one each day and avoid the whole rice debacle. If I find a job, I can live here.

But... my Mom is getting re-married soon, and I'm really missing her, Sandy, Katie, and all my friends. I'm so close to home it feels *weird* to just stay here. I could just lie on the beach for a week and go home. *Yep, beach it is.*

I'm on Waikiki, lounging in the sun. It feels strange to have to keep my bikini top on now. It's strange to hear English with a Yankee accent.

I notice a large group playing volleyball. They're all laughing and having a great time, and I overhear distinct New York accents. A twinge of loneliness washes over me because they look like a fun group of friends. At that moment, the ball comes rolling over to my towel. A brawny, muscle-y guy comes over to retrieve it and asks if I want to play as well.

Did I just make that happen? Of course, I say yes.

It turns out I was right. Friends and couples from New York, all cops and firemen. There are six girls but seven guys, and my muscled friend is Bob. (Bob Midnight? Spirit Guide Bob?) He's tired of being the odd man out and asks me to join them all for dinner tonight.

I tell him I'm returning from a world-wide adventure and have no money left. He smiles, he insists it's his treat, that he'd love to hear about my adventures. *Is this a date?* I'm not sure.

I arrive at their fancy hotel dressed in my best new Bali sundress. Everyone's in the lounge and Bob jumps up to walk me in when he sees me. *Date?* He re-introduces me to everyone and sits on a different couch from me. *Not a date.* He gets me a drink. *Date?* He talks closely with one of the other girls in the group. *Not.* The other people are interested in hearing my stories about Africa, Indonesia, and New Zealand. He holds my hand while we walk to the dinner table. *Date.* He pulls out my chair. *Date.* He sits across from me and laughs at my jokes. *Date.*

Finally, he tells me he has a live-in girlfriend who's also a firefighter back home and she couldn't make the trip last minute, but these are all her friends here too. *So, not a date.* He says he's just been feeling like the 13th wheel, and would I join them for dancing and breakfast tomorrow. I tell him my literal budget of $8 and he laughs, telling me everyone is enjoying my stories, my company. He insists and I accept.

Each day I meet Bob with his group and we spend the day together on the beach, playing volleyball, snorkeling, watching the sunset, swimming at their pool, eating glorious breakfast buffets, lunch buffets, and dressing for fancy fine dining dinners, even a luau with a roasted pig from the ground and a live hula dancing show. We all spend each night dancing the night away in the beachside night clubs. After each meal Bob and his friends invite me to the next. I play coy until they insist... this goes on all week, and I love it.

Unfortunately, I have only a couple of nice sundresses, so they get to see them over and over. I try to wear my hair differently, so maybe they won't notice.

They're leaving just two hours before me to fly back to New York. I have the greatest week in Hawaii, strictly because of their laughter and kindness to a weary solo traveler.

*****

Today's the day.

I've taken the time to dress in new gauzy white pants with a white crochet tank top, my custom deep maroon suede jacket and matching high heel slides. Lots of mascara and lipstick with my (*it's still too short, dammit*) curly hair. In the airport, I buy a 95-cent postcard and realize I've just spent eight days in Hawaii with $8, so I'm heading home with $7 in my pocket.

The wheels lift. The plane steadies. I put my Walkman on with the first tape I listened to flying into South Africa. The soundtrack of '*Natural Born Killers.*' Leonard Cohen and Cowboy Junkies are making me feel nostalgic for each place I've now called home, places I'll forever cherish in my heart throughout my life.

I'm so excited to see my family again. I just hope it's not my *only* overseas adventure, ever.

California

Los Angeles

- L.A. Woman -
The Doors

The raven hair, the sapphire eyes. There's my mother's face!

I fall into her hug that utterly feels like home. Like I'm a five-year-old girl who fell out of the tree I was climbing. And my mother's hug washes away all the pain, all the loneliness, all the exhaustion.

As I pull back, I'm now the adult. She can see in my eyes I have so much, too much, to share. Things no one will understand, unspoken adventures that need no words for now. I know in this moment, because of her own travels, she fully 'gets' me.

I've been home for three days, and with a borrowed $20 dollars from Mom, I decide to stop into the grocery store for a few things. The aisles are filled, I mean FILLED with cereal boxes, peanut butter jars, and shampoos. And everything is way cheaper than I'd been paying. I'm confused why we, as a society, need so much stuff, yet truly grateful that I have thousands of things to choose from in one store. I've not seen a supermarket like this anywhere, even the western countries.

Is this the reward of capitalism? No, not crony-corporation-capitalism, real free market driven capitalism. If I want a job tomorrow, I can go get one. From my own choices, I can study the skills I need, or work my way up to earn the lifestyle I want.

I meander next door to Best Buy to look at CDs and see what I've been missing out on. I've already mixed-up slang *(That's the Bomb!)* thinking it was a bad thing.

I'm reaching for a new '*311*' CD when I hear a deep, sexy, familiar voice.

"Well. And then there's Pamela," he says.

I look up over the music racks, instantly melting into a pair of dusty jade green eyes.

A tall, golden Greek God of a blond Harrison-Ford-face of an old boyfriend.

He'd joined the army and left after we shared a few dates years ago. There had always been a flirty electric energy between us. *Unfinished business? Could the Universe be so kind?*

"Well, hello handsome! If it isn't Dan the Man!" I smile. We walk to the end of the aisle and hug tightly as my heart skips a beat. I think I could die here in these arms. I can't stop grinning. I just blurt out "Are you single?"

His face clouds over a bit, letting go of the hug, "No, I, aww... I'm living with someone..."

I frown and wave my hand, callously, shamelessly, telling him I don't even feel like catching up then. We laugh and part ways.

In the next few days, still the proud owner of just $7, I've landed a waitress job at a steakhouse. I seem to have a bit of an accent, a mixture from the many places I've been, causing constant questions from customers about where I'm from. A lot of our regulars are old friends of mine, including some stuntmen in the movie business. One of them hears about my journey and pulls me aside.

"This must feel so weird now, this everyday life compared with what you've just done, the things you must've seen...?" he says.

"It does feel surreal," I smile, he understands me. I realize, for the first time, that I'm just not enjoying my life, or living to the fullest now that I'm home. I'm just going through the motions.

"I want you to know, it was truly brave and courageous of you to go out and do this all on your own." He holds my shoulder like he's proud of me. Didn't he just jump from a 20-story building for a shoot, yet *I'm* brave?

He tells me he's been all over the world for filming and—same thing—no one cares. "I want you to know you're a rare breed, and any man would be lucky to have a girl like you. Don't ever let it hurt your feelings if people just don't care about your journey, they just don't get it."

"It *has* felt strange," I nod in agreement. "When I say the last stop was Hawaii, they light up and say '*You went to Hawaii? How was it?*' as if that's not America as well. It's kind of '*Twilight Zone*' time!" We laugh. It's good to hear I'm not alone. Or crazy. Mostly.

*****

After celebrating my mother's gorgeous wedding on the Queen Mary ship, I really do try to settle in.

I. Cannot. Settle. In. After my accidental years abroad, I simply no longer fit.

My sister and I decide to spend time traveling together again. Why not a few weeks in England to visit some cousins?

I make myself a turkey and Swiss cheese sandwich with spicy mustard, grab a can of root beer, along with a mug filled with Butter Pecan ice cream. Still can't get over how much food is in the cupboards at my own home.

I start calling my new English overseas friends.

I call Heater first because I know she's returned home from Australia by now. She'd love to see us, we can stay with her, but she's also going with her boyfriend to the Canary Islands, off of Morocco, around that same time... why don't I fly in early, join her there, then join Sandy in England?

My eyes twinkle. I scarf my sandwich while we're planning.

I call Skyler next, because she's also back home from New Zealand, and she'd love to see us, and would I like to go to Paris with her while I'm there?

Gulp, yes, please! I drink my root beer, then get smart, pouring the rest of the can over the ice cream, creating a root beer float. I grab a huge spoon so I'm slurping while we're organizing.

Next, I call Lucy, home from Indonesia, who's not going to be in England. But tall Sean from Bali is now living in London, and the guys Weston and Perry are back home. I could see all of them. She mentions she'll be in Los Angeles a few weeks after I'd get back from Europe.

"Pammy, I'm going through the South Pacific Islands to study Eco-Tourism," she tells me.

"What's Eco-Tourism?" I ask, clearly envious.

"It's a new thing where it's a much more Earth-friendly way to travel. After LA, I'll be in Easter Island, Tahiti, Moorea, Cook Islands, Fiji, Western Samoa, Tonga, New Zealand, and Australia... wanna come with?"

It hits me.

I understand.

These are the people I've met...

These are the adventures I've had...

This is whom I've become...

I catch a glimpse of my reflection in the back of my large spoon.

I still look like me... *only, wilder somehow... untamed forever.*

I steady the dirty dishes in my hands. A huge grin spreads across my face.

"Yes."

I'm lost in my daydream...

- Good Riddance -
(Time of Your Life)

Green Day

Finale

"Me? In a video?" I asked.

"Yes, would you like to be in my music video?" he tried again.

I smiled. "No," I giggled, shifting the dirty dishes in my hands as I laid their check on the table.

"No?" he raised his eyebrows, genuinely surprised.

"No, but if you'd like to be a waiter here, I can put in a good word for you to the management. You might have to start out as a busboy, but you could work your way up in no time," I joked, with a wink.

His raised eyebrows slowly sank, and he didn't smile back as my joke fell flat. "Oh. Okay. Nevermind, then." He gave me an odd look as I went off into the kitchen to deliver more food.

It was the odd look on his face after my lame joke. I doubted myself. Oh well, I still had another chance to check him out properly after I got the credit card. I'd explain myself. We'd have a laugh about it, and I'd be in the video after all.

"I think it *is* Rod Stewart," said my coworker, carrying drinks away.

I gave myself a quick touch up of lipstick and headed back over to the table. As soon as I rounded the corner, I saw the busboy clearing the table.

They were gone.

My first famous encounter couldn't end like this.

I sprinted up to the hostess. She had the paid bill in hand, it said *Rod Stewart* on the credit card slip. I could still see his spiky hair as he climbed into his fancy car in the parking lot.

I meant Yes! I'll be in your video! Rewind Rod!

Ask me again! YES, yes, *y-e-s...*

A grand missed opportunity. An unnoticed turning point in life. Why was I suspicious? I should've trusted him.

*****

Months later, I was at home with VH1 on in the background. A new video premiered by, you guessed it, Rod Stewart. There was a video girl who looked pretty much like me.

My mom laughed, "Oh, Pam, of course Rod Stewart would remember you. You're the only waitress in the history of Los Angeles who turned down a part on film!" She gave me a big squeeze.

Thanks, Mom...

Forevermore, my answer is YES. To EVERYTHING.

Thank you for changing my life, Rod Stewart.

- In Rod We Trust -

Special Thanks

If you need the sea in Newport Beach, California, USA -
J.J. is now a tourist boat captain. He began this journey with me in Africa and encouraged me to keep going solo. Check out his adventures in his own memoir titled:
"The World's Richest Busboy" by J.J. Brito

If you're partying in Lake Taupo, New Zealand -
DJ Paul is on the radio at **Lake FM 89.6** so listen in for obnoxious shenanigans over the airwaves.
Mr. Toad is back to the scene of the crime— When in ROAM's cover— after 13 years at the Macau Tower in Hong Kong, he's moved home to Taupo. If you need a wild bungee jump, he promises he won't push!
A.J. Hackett Bungee

If you're traveling in the Netherlands -
J.R. will take you for a skydive if you're feeling adventurous— in every way. For lots of laughs and excitement, check him out at:
Beyond Cool Skydive *Texel BV*

And my warmest additional gratitude to...

Katie Kray Reilly and family, Tobie Jenkins, Deborah Vouaux, John Characky, Cindy Nelson Roberts, Jean Groskreutz, Dea Fischer, Millie Rose, Pam Carrie, Jennifer Lynn, Jodi Harari, Brian David, Sal D'Elia, and Sohaila at SohailaInternational.com...
for answering all my stupid questions.

And Thanks to YOU!
I sure do hope you enjoyed my adventure...

As a new author, it truly helps me if you
PLEASE LEAVE A BOOK REVIEW ONLINE!

From the Peanut Gallery

"Allegedly...!"

"But... *why go?*"

"Weren't you scared?"

"Wasn't it dangerous?"

"I don't think I could do that."

"You're piss funny from *Go to Whoa!*"

"Wait, it looks like he pushed you!"

"What were you running away from?"

"Did you get it all out of your system?"

"I laughed so hard I spit coffee onto my dog!"

"I'm in the book? Oh no! Should I go into hiding?"

"Shoving is such a harsh word, I prefer '*assisting*'..."

"You're like trying to sip water from a rushing fire hydrant!"

...and my personal favorite:

"If someone doesn't like your book, you tell them your Mom says it's the best book out there and they need to get it together!"

- In Mom We Trust -

About the Author

Pamdiana Jones lives in Laguna Beach, California, USA, with her beloved husband Daniel. Yep, the tall, old boyfriend, golden Greek God of that blond, Harrison-Ford-face, owner of those gorgeous dusty jade green eyes.

They are parents to the happiest, funniest, smartest, kindest, most beautiful twins on planet Earth.

Pam continues to write her next two travel memoirs and is creating a children's adventure book series.

She hopes to continue making you think and making you laugh.

She'd love to hear from you!

Pam@PamdianaJones.com

www.PamdianaJones.com

Facebook-Instagram-Pinterest @ PamdianaJones